The Implementation of China's Science and Technology Policy

The Implementation of China's Science and Technology Policy

Q. Y. Yu

Foreword by Edward E. David, Jr.

QUORUM BOOKS
Westport, Connecticut • London

Library of Congress Cataloging-in-Publication Data

Yu, Q. Y., 1931–
The implementation of China's science and technology policy / Q.Y. Yu ; foreword by Edward E. David, Jr.
p. cm.
Includes bibliographical references and index.
ISBN 1–56720–332–9 (alk. paper)
1. Science and state—China. 2. Technology and state—China.
I. Title.
Q127.C5Y8 1999
338.95106—dc21 99–21588

British Library Cataloguing in Publication Data is available.

Library of Congress Catalog Card Number: 99–21588
ISBN: 1–56720–332–9

First published in 1999

Quorum Books, 88 Post Road West, Westport, CT 06881
An imprint of Greenwood Publishing Group, Inc.
www.quorumbooks.com

Printed in the United States of America

The paper used in this book complies with the Permanent Paper Standard issued by the National Information Standards Organization (Z39.48–1984).

10 9 8 7 6 5 4 3 2 1

I wish to extend my sincere appreciation to Dr. Edward E. David, Dr. John W. Garver, Dr. John E. Endicott, Dr. John R. McIntyre and Dr. Mark J. Ballam. Without their encouragement and help, I could not have presented this book to its audience so far away from my home country. I am grateful to the Sam Nunn School of International Affairs and the GT Center for International Business Education and Research at the Georgia Institute of Technology; their valuable support and assistance made possible the finalization of the manuscript of this book. Also, I wish to express my thankfulness to my wife, Xun Dezheng, for her support and assistance.

Contents

Tables ix

Foreword by Edward E. David, Jr. xi

Introduction xiii

1. China's Scientific and Technological Development During the Period from 1949 to 1978 1
2. An Overview of the General Orientation and Guidelines of China's Science and Technology Policy 8
3. The Reform of the Management System of Science and Technology 14
4. The Reform of the Appropriation System 18
5. The Development of a Technology Market 22
6. Strengthening the Capability of Technology Development and Assimilation of Industrial Enterprises 30
7. Reform of the Scientific and Technical Personnel Management System 36
8. China's Opening and Technology Acquisition Policy 42

9. Readjustment of Science and Technology Policy in the 1990s 65

10. China's Industrial Policy and Technology Acquisition 75

11. The Development of Agricultural Science and Technology 85

12. Township Industry and the Development of Appropriate Technology: The Spark Program 104

13. Policy on the Development of Infrastructure Industries 121

14. "Mainstay Industries" and Technology Policy 139

15. Emphasis on Basic Research 166

16. High-Tech Research 172

17. The Torch Program and the Progress of High-Tech Industry 184

18. Non-governmental Science and Technology Enterprises 199

19. Policy on Sustainable Development 208

20. Epilogue 222

References 225

Index 229

Tables

1. The Growth of the Amount of Technology Trade Between 1983 and 1989 23
2. Development of New Institutions Devoted to Technology Development and Trade 24
3. Composition of Different Kinds of Technology Contracts 26
4. Technology Purchasing and Selling by Large and Medium-Sized Enterprises in 1987 31
5. Sales of State-Owned Enterprises 32
6. State-Owned Large and Medium-Sized Industries 34
7. Different Forms of Integration of Research Institutions (above County Level) with Enterprises 39
8. Output of Various Durable Home Appliance Products in 1994 48
9. The Growth of Foreign Investment in China 51
10. Growth of Import and Export of Enterprises with Foreign Investment 52
11. Survey on 1,180 Sets of Equipment Owned by Large and Medium-Sized Enterprises 67

12. General Situation of Township Industry by the End of 1995 105
13. Achievements of the Spark Program 115
14. Composition of Investment of the Spark Program's Projects 116
15. Development of the Railway System in China 130
16. Development of the Highway System in China, 1992–1996 131
17. Development of Water Transportation in China 132
18. Passenger Turnover of Civil Aviation in China, 1992–1996 133
19. Growth of Telephone Spread Rate in China, 1978–1996 135
20. Growth of the Chemical Fertilizer Industry, 1992–1996 150
21. Growth of Synthetic Fibers Production, 1992–1995 151
22. Criteria of Economic Scale of Fixed Asset Investment Projects 152
23. Growth of the Automobile Industry, 1993–1995 161
24. Transformation Cycle of R&D Results: Statistics of 54 Companies 205
25. Sewage Displacement Throughout China, 1990 and 1993 213
26. Waste Gas Displacement in China 216

Foreword

For many centuries, China has had a history of technological progress. Well known are its inventions of gunpowder, pyrotechnics, seismic sensors, the magnetic compass, papermaking, movable type, fine porcelain and silk processing and weaving. In more recent times China has innovated further in technology (the chemical synthesis of insulin during the 1970s was a landmark) and also has become aware of science as a driver of technology. Until the establishment of the People's Republic of China in 1949, China's attention was focused primarily on details, the nuts and bolts of technology and instrumentation. Only since that time has the idea of national policies for science and technology as a means for economic development become a priority. The genesis and progress of such policies is the subject of this book by Q. Y. Yu. It is a look from the inside at events leading to China's emergence as a world player in both science and technology.

Mr. Yu was my personal host on two trips to China. He accompanied me on visits to the far reaches of China. Mr. Yu is a cultured, observant and astute person. He understands the importance of knowledge and scholarly research to the well-being of China. He understands, too, the connection between research and economic development. His wisdom in these matters shows through clearly in his writings and commentary on events in China.

Mr. Yu has extraordinary dimensions, especially his concern for people. He has visited the United States many times and has been an invited lecturer at the Georgia Institute of Technology. His thinking about science policy and technological innovation has been broadly and well received.

I highly recommend *The Implementation of China's Science and Technology Policy* to all who visit China and who value an understanding of its ways and its institutions. The book is especially valuable for scientists, engineers, diplomats and businesspeople. It will become an essential part of English language literature on China.

Edward E. David, Jr.

Introduction

China has made great accomplishments in the past 50 years since the founding of the People's Republic. However, she has pursued an uneven and rough path in this process. One may say the same with regard to scientific and technological development.

This book will give an overview of China's science and technology policy as well as its implementation. China began to publish official documents addressing science and technology policy in the mid-1980s. During the period of time before 1966 (the beginning of the Cultural Revolution), under a highly centralized planned economy, the guiding thought and focuses of scientific and technological development had been included in various Party and government documents. These served as administrative orders issued to the government functionaries. During the period of the 10 years of Cultural Revolution, the fundamental policy was to destroy the "bourgeois intellectuals" controlling science and technology. In fact, what the "revolution" destroyed was science and technology itself. It was not until the late 1970s and early 1980s, when China reopened itself to the outside world, that the government publicly put forward principles and policies addressing the development of science and technology. This effort necessarily involved a large number of scientists and engineers as well as government functionaries in determining the scientific and technological structure and mechanism. During the mid-1980s, a systematic science and technology policy had been shaped, and with further economic reform, science and technology policy was further developed. The main content of this book gives a detailed description of the science and technology policy as well as its implementation in the 1980s and 1990s. It will begin, however, with a short account of what happened before 1978, since the significance of what happened after that decisive year can be adequately understood only in the context of what had gone before.

It is difficult to give a stand-alone description of the history of China's science and technology. What has happened in the realm of science and technology is inseparable from politics and economy. The impact of the political and economic situation on the development of science and technology is similar for many countries, but that influence has been stronger in China. The influences of the political and economic changes in the past half century have been unprecedented in both depth and scope.

When the Republic was founded in 1949, only a poor legacy in science and technology policy existed. China had been a country with a glorious tradition in the development of science and technology in ancient times. However, in recent history, China's closed door policy isolated her completely from the industrial development in the West and left her ignorant of the rapid economic growth in the outside world. It was as late as the mid-nineteenth century, after China's closed door was broken open by the gunboats of the Western powers in the Opium War, that progressive Chinese intellectuals were awakened from the fond dream that China was still the most powerful country in the world in all respects. The first thing they learned was that the weapons of the Western powers were indeed more powerful than those of China. The progressive intellectuals, as the most sensitive group of people during the time, began to open their eyes to the outside world. It was no wonder that they started with geographical studies, followed up with studies of political and military systems, and finally turned to technological development.

The learning process broadened the vision of this group of intellectuals, which included quite a few high-ranking officials with political power. A new concept gave rise to the question: Why shouldn't we learn skills from the foreigners first and then bring them under our control? A new school emerged under the name "foreign affairs school," which consisted of a group of visionary high-ranking officials who were involved in foreign affairs. The doctrine of the school was: Chinese tradition should be taken as principal, and Western learning should be employed for utilization only. Under the doctrine, young students were sent to study abroad, with the first batch of students sent to the United States. During the early stage of sending students abroad, the focus was on naval training with the purpose of building a Chinese naval force. China's first naval force, the Beiyang Fleet, was established with most of its officers trained in Britain. It was interesting that the first monograph of the Western scientist T. H. Huxley, *Man's Place in Nature,* was translated into Chinese by an outstanding writer who was also one of the students studying naval matters in Britain. The translation was widely read by progressive intellectuals during that time.

The practice of "learning from the foreigners" started with a military purpose (buying foreign weapons and sending students abroad) and followed up with efforts to build up China's own modern industry. The development of modern industry certainly required the study of modern industrial technology. The century-long arduous path of the development of China's modern science and technology had begun.

Industrial and social development required the establishment of a modern educational system. The first university in China, Capital University (later renamed Beijing University), was established in Beijing 100 years ago. However, during that time, Chinese traditional culture was still orthodox, and modern science and technology only played the part of a sort of ad hoc skill. It took decades to break a path through the obstacles of traditional culture to pave the way for the development of modern science and technology. It was as late as the first quarter of the twentieth century that modern physics as a discipline was firmly established in the higher education system in China.

When looking back to the early development of science and technology in China, it is clear that it was owing to a sense of national crisis and the desire to defend national tradition that China started her efforts to develop modern science and technology. In the process, many Chinese scientists felt that they had an obligation to work for development, as it was a way to help the nation become stronger. Therefore, right in China's history, devotion to science stemmed from the sense of national salvation. Many of the scientists of the older generation were stepping into the new People's Republic with this mentality to continue their efforts along the arduous path to prosperity.

The Implementation of China's Science and Technology Policy

1

China's Scientific and Technological Development During the Period from 1949 to 1978

A new prospect for scientific and technological development arose when the new Republic was founded in China in 1949. As the country was just coming out of war and concentrating her efforts on the restoration of a national economy, it was impossible for her to expend much of her resources on expensive scientific programs. The efforts made in respect to science and technology were focused on reorganizing the existing scientific undertakings and science manpower. The legacy left by the old China in terms of science and technology was very limited. Before the founding of the People's Republic, there were only 180,000 university graduates, 50,000 scientific and technical personnel, 40 scientific research institutes and a few hundred research scientists. As early as 1949, the Chinese Academy of Sciences was established to devote itself to China's basic research and major applied research. In 1952, when the period of restoration of the national economy was concluded and the first five year plan started, China began to invest substantial resources in the development of science and technology and started her efforts to realize national industrialization.

National industrialization had been a century-long dream of the Chinese people. However, it turned out to be a much longer and tougher process than originally expected. Since the officials of the "foreign affairs school" first set out to develop national industry, almost a century had passed, with achievement limited to the sector of light industry. Heavy industry had been almost completely neglected. With a view to quickening the pace of industrialization, in the early 1950s, 156 large industrial projects were introduced by the Soviet Union. These projects covered the major industrial departments of heavy industry in an attempt to form China's initial industrial system. During the 1950s, large numbers of Soviet experts worked in China to help with installment, operation and management. Since China's initial industrial system was established with technology, equipment and a management system introduced from the Soviet Union, China's industrial development couldn't help but closely follow the Soviet pattern for a fairly long period of time. The influence of the old Soviet system can be found even today. The industrial support of the Soviet Union in the 1950s had both positive and negative sides. It did play an important role in the formation of the initial industrial system in China and help China to take a big leap in the development of industrial technology. However, the rigid and inefficient Soviet management and operation system, as well as the slower technological progress derived from an impotent industrial research and development (R&D) system under a central planning economy, has been a heavy burden for China's industrial development.

In 1952, with a view to preparing for the coming high tide of socialist economic construction, the reorganization of the educational system was launched with the closing of missionary and private universities. Many universities, colleges and disciplines were readjusted and reorganized. The curricula were rearranged generally, so as to gear education to the needs of socialist economic construction.

A central planning economic system was adopted by China right from the outset of economic construction, based on the belief that it was the ideal operational system for a socialist economy. The Soviet Union, being the first socialist country in the world, with several decades of experience in implementing a planned economy, was certainly the example for China to follow. Therefore, "learning from the Soviet Union" became the principle in economic construction as well as in science and education. The new scientific research system, of course, followed that of the Soviet Union. Just as in the Soviet Union, China's scientific research was part of central planning. The State Science and Technology Commission was set up and placed in charge of the overall planning of scientific research programs. This commission allocated budgets for scientific and technological research. For a period of 17 years, from 1949 to 1966, China's scientific and technological force grew rapidly. Though the starting point was low, a number of scientists at senior levels studying and working abroad came back to the country in the early 1950s, and in addition a large number of scientists and engineers of a new generation were emerging. By 1952, when the first five year plan started, there were 428,000 scientific and technical personnel throughout the country. By

1966, just before the Cultural Revolution, the number soared to 2,458,000. Of course, the size of the scientific and technological force was still rather small, especially relative to the large population of the country. However, taking advantage of the highly centralized administrative power and the central planning economy, the central government was able to concentrate a scientific research force to tackle the few most urgent and difficult issues and bring about comparatively quick successes. For instance, China succeeded in launching her first large rocket in 1959, made a successful experiment on the first atomic bomb in 1964 and in 1965 accomplished the first totally synthetic crystalline bovine insulin in the world.

Nonetheless, the development of industrial technology had been slow and sometime even stagnant. This was mainly due to the rigidity of central planning. The development of industrial technology is inseparable from the industrial and agricultural systems. Under the centrally planned economy, the industrial research system, instead of being integrated, was parallel to industrial production. Under the centrally planned economy, all State-owned industries were directly under the administrative control of the government departments. There were two different levels: the industries subordinated to central ministries and those subordinated to local government departments. All aspects of industrial production were controlled by related government departments. Each industry obtained its raw materials from government allocation, manufactured its products according to the quota set up by the government department and transferred its end products to the government distribution system. Consequently, no effort needed to be made by the industry in terms of procurement of raw materials or marketing of products. The industry was responsible only for production, or, in other words, to satisfy its assigned quota. The industry didn't need to worry about economic benefits or competition, or about technological renovation or innovation. What enterprises faced was not the market, but administrative orders given by superior government departments. As a result, industry didn't need to take any initiative. Under this situation, it was very difficult for industry to renew products or processes.

A typical example was the automobile industry. China started her automobile industry in the 1950s. The First Automobile Factory in Jilin Province in China's northeast was one of the large industrial projects aided by the Soviet Union. The factory mainly produced trucks, but in the late 1950s it turned out its first limousine. That limousine remained unchanged for 20 years. Such a phenomenon has been called "a system of twenty year consistency." Development research, instead of being subordinated to industry, was put under government departments, subordinated to various industrial ministries. At the central level, the ministries set up their own research institutes for development research.

Plans for improvements in products and processes were decided by the government departments, and the designs were worked out by the research institutes. Only when the blueprint for a prototype was ready would it be handed over to the factory whose job was to turn out the new product or process. Such a system couldn't but bring about the divorce of research from production. People called such a system the "government directed research system." During the post-1978 period, China's reformers tried hard to transform the "government directed research system" into the "enterprise directed research system."

Even under a rigid centrally planned economic system, as China had the opportunity, for the first time in more than a century, to construct the country in a peaceful environment under a unified, highly centralized authority and organize the masses of people on an unprecedentedly large scale, the first five year plan was completed and turned out to be success. Economic development gave impetus to the further development of science and technology. By 1955, scientific and technological research institutions numbered more than 840; scientific and technical personnel had increased to more than 400,000.

In 1956, at the high tide of economic construction, a national conference on intellectuals was held, during which Zhou Enlai indicated the important role of science and technology in socialist construction and issued the call for "marching toward science." Subsequently, the State Council established a Science Planning Committee made up of more than 600 scientists and experts. This committee was charged with formulating China's first long-range plan for scientific and technological development (i.e., "National Plan on the Prospect of the Development of Science and Technology [1956–1967]"). In view of the requirements of economic development and national defense, the plan listed 57 tasks, including a series of tasks relating to basic, applied and development research. The most important step was to adopt six urgent measures, to develop computer, semiconductor, automation, radio, nuclear and jet technology. This gave birth to a series of emergent technologies in China and laid the foundation for the growth of a number of new industrial departments. A course for China's scientific and technological development was outlined in the long-range plan. The gap between China and the world in respect to science and technology had begun to narrow.

Nonetheless, the even and "slower" pace of the first five year plan didn't satisfy the more ambitious leftists. In 1958, the Great Leap Forward was launched. The next year an anti-rightist campaign consolidated the frantic leftist trend of thought in pushing forward economic construction at unbelievable speed. Goals were set on an impractical basis, which couldn't help but encourage boasting and exaggeration. Blindness in issuing administrative commands and an eagerness for quick success hastened the pace of economic development but created a chaotic situation in economic construction. This was especially harmful to the scientific work, as results in scientific and technological research could only be the fruits of accuracy and patient work.

The Great Leap Forward was eventually abandoned as the economy collapsed. In order to restore the normal order, which had been sabotaged by the Great Leap Forward, scientific and technological work was readjusted, and a document on the readjustment was worked out jointly by the State Science and Technology Commission and the Chinese Academy of Sciences, entitled "Fourteen Points of Opinion on the Current Work of the Natural Science Research Undertakings," in the early 1960s. The document indicated that various kinds of leftist trends should be rectified and policies regarding scientific and technological work should be clarified. The most notable issue put forward by the document was that the "fundamental task" of scientific research undertakings was to achieve results and bring up people of talent. In 1962, during two conferences in Guangzhou, Zhou Enlai and other leaders reiterated that China's intellectuals, like workers and peasants, comprised one group of China's laboring people. Since the anti-rightist campaign in 1957,[1] intellectuals were treated as "bourgeoisie intellectuals," as their outlook was supposed to be of the category of the "bourgeoisie." The new announcement by the leaders actually relieved the intellectuals of the label of "bourgeois intellectuals." These measures helped bring scientific and technological work back to a normal track. Under this situation, the State Science and Technology Commission formulated the "Scientific and Technological Development Plan for the Period from 1963 to 1972." It was an attempt to push forward the development of science and technology along the original track set up in the early 1950s.

Barely had the situation been regularized when the Cultural Revolution began in 1966. This began a 10-year period during which Chinese scientific and technological efforts were seriously sabotaged. The Cultural Revolution was not just a repetition of the leftist trends of the late 1950s. The scientific and technological work of the entire 17 years before the Cultural Revolution was negated and repudiated as a "reactionary revisionist line." Research institutions were dissolved, research facilities and science and technology materials were destroyed, experiment bases were sabotaged, the regular working order was disturbed and the entire scientific and technological apparatus was broken down and paralyzed. A large number of scientific and technical personnel suffered from persecution. Quite a number of veteran scientists and experts were labeled "reactionary authorities," and the scientific and technological force created since the founding of the People's Republic was virtually destroyed. During the Cultural Revolution, it was strictly prohibited to learn from the advanced experiences of foreign countries in respect to science and technology. Any effort to learn from foreign advanced experiences was repudiated as "slavish comprador philosophy" or "crawlism." During this time, what prevailed was a closed door policy that isolated the country from the outside world and broadened the gaps in science and technology between China and the world's advanced levels.

In addition, the educational system was severely injured by the Cultural Revolution. Even high school classes were ended for a few years, as students were mobilized to serve as "red guards," the motivating force of the Cultural Revolution. But as the inexperienced youth were instigated behind the scenes by political factions, they split into different groups, launched into ferocious fighting against each other and caused a turbulent and chaotic situation in society. With a view to quieting the turbulence, beginning in 1968, millions of youth were sent to the countryside to take part in farming. When most of them came back from the countryside after the Cultural Revolution, they had to restart their education where they had left off or find jobs. From the standpoint of science and education, China had lost a great deal. The Cultural Revolution had a far-reaching influence. In the late 1980s and early 1990s, when most university graduates of the 1950s and early 1960s began to reach retirement age, it was very difficult for the university and research institutions to find qualified individuals to take over their responsibilities as faculty and researchers.

When looking back on the history of the 27 years from 1949 to 1976, it can be seen that all through that long period, an intense struggle was under way between efforts to build up a scientific and educational system suited to a planned economy and an ever stronger leftist force that sought to break up these efforts by putting the top priority on politics and revolution.

During the two years after the Cultural Revolution, much effort went into theoretical criticism and repudiation of the practice of the Cultural Revolution. The Cultural Revolution had brought heavy calamities to China's educational system. Since the very beginning of the Cultural Revolution, the educational front was taken as an area controlled by "bourgeois intellectuals." Quite a number of faculty members of universities and research institutes had been persecuted, their positions and reputation not restored until the conclusion of the Cultural Revolution. Right after the Cultural Revolution, the most important issue was how to clarify China's guiding thought and bring order out of chaos. Initially, the restored "order" was supposed to be the planned economy, which had been overturned by the Cultural Revolution. In 1978, at the Third Plenary Session of the Central Committee of the Eleventh National Congress of the Communist Party, the policy of reform and opening was officially put forward. In 1979 the first National Science Conference was held. It announced that intellectuals were part of the working class and made important contributions to the construction of China's socialist modernization. The intellectuals involved in scientific and technological work were greatly encouraged and began to speak with strong feeling at long last: "Welcome the spring of science."

NOTE

1. In 1957, the anti-rightist campaign was focused on criticizing and repudiating intellectuals with remarks that belonged to "rightist" trend. In 1959, the anti-rightist campaign was an inner party struggle that focused on opposing "rightist opportunists." Most of the victims of these anti-rightist campaigns were redressed after the Cultural Revolution.

2

An Overview of the General Orientation and Guidelines of China's Science and Technology Policy

The year 1978 was of major importance in China. It was at the Third Plenary Session of the Central Committee of the Eleventh Congress of the Communist Party of China that the decision on the construction of the four modernizations (agriculture, industry, defense and science and technology) was made. It was declared that in order to realize the four modernizations, China should shift to taking socialist economic construction instead of class struggle as the central role and to carrying out reform and opening policy. As a development goal, it was necessary to quadruple the annual national output value by the year 2000.

Science and technology was taken as one of the key elements in realizing the development goal. At the First National Science Congress, held in 1978, Deng Xiaoping pointed out, "In four modernizations, the modernization of science and technology is the key. Without modern science and technology, it will be impossible to construct modern agriculture, modern industry and modern defense, and it will be impossible to have the development of national economy at high speed." At the Congress, Deng put forward two major issues: science and technology is the force of production (later on, it was further emphasized as "first force of production"[1]), and intellectuals are part of the working class. Intellectuals were once again relieved of the label of "bourgeois intellectuals" and granted the status of "working class intellectuals," which they had longed for for so many years. In 1982, a general orientation for China's science and technology policy was issued by the government: "Economic construction must rely on science and technology, and science and technology must be oriented toward economic construction." When specifying this new orientation, the leader of the State Council at that time pointed out that in 1980, the total value of China's industrial and agricultural output was 710 billion yuan; by the year 2000, it should be brought up to 2,800 billion yuan, under the premise of constant improvement of economic efficiency. Science and technology should play an important part in the realization of this goal, which was explained in detail:

> How are the conditions of our financial resources, energy resources and materials? Is it possible for these conditions to ensure the realization of the goal? To approach from the conditions of capital, energy and materials, we may probably have a concept that it may be possible and may not. Because, if calculating according to the current technical economic indexes, we are seriously short of capital, energy and materials. However, if, in the process of the construction during the coming two decades, various departments of our national economy can be gradually shifted onto the basis of new technology and various technical economic indexes can be brought up to a brand new level, then, the conditions of capital, energy and materials will be able to ensure the quadruplicating of the total agricultural and industrial output.

The situation of energy resources and materials was taken as examples. In 1980, various kinds of energy produced in China, converted into standard coal, were equivalent to 630 million tons. The total agricultural and industrial output value of 710 billion yuan was realized with the 630 million tons of fuel and power. By the end of the century, the energy supply might be doubled. Could the total output value be quadrupled by the doubled energy supply? It might be or might not. If people just stayed with the current technological basis, based on the current ratio between total output value and energy, China could only double the total output value. However, on the basis of advanced technology, if people could make the ratio between total output value and energy rise to a new level, the situation would be different. To approach the unit energy consumption of total national output value, China was lagging behind even as a developing country, let alone in comparison with developed countries. At the time, Japan consumed about the same amount of energy as China did, roughly equivalent to 600 billion tons of standard coal, but its total national output value was three or four times China's. Of course, some elements were incomparable, such as a different energy composition and a different product mix, but the gap was still too big, mainly owing to Japan's higher technological level. By the end of the century, China might not be able to reach a level that high, but China should strive to come up to a level that would realize the total output value of

2,800 billion yuan, with 1.2 billion tons of standard coal. It would not be enough to rely on the breakthrough of a few enterprises or a single industry department in technological renovation. Equipment with a high energy consumption, used by all the industries, should be renewed; even so, it would not be enough. All industries needed to carry out a series of technological renovations to change their techniques, process and product mix. Only in this way could energy consumption be brought down by a big margin. Therefore, the shortage of energy should be made up by science and technology. A similar situation existed in the field of materials. People should come to science and technology to pursue cement, metallurgical materials, synthetic materials and various kinds of new materials and substitute materials. In these circumstances, it was essential for China to closely link up the development of science and technology with economic construction.

The "Outline of the Report on the Guidelines for China's Scientific and Technological Development,"[2] submitted by the State Science and Technology Commission and announced by the Central Committee of the Communist Party of China and the State Council, put forward five guidelines in the light of this general orientation:

1. Science and technology should be developed in coordination with economy and society, and the promotion of economic development should be taken as the task of highest importance for science and technology.

2. Research on industrial technology should be strengthened and technologies should be correctly selected for the formation of a rational technical setup.

As China has limited resources for scientific research, it is most important to select technologies carefully. The Report Outline lays special emphasis on research on the technology of production. In the past more than three decades, China did have a successful experience in concentrating her limited research force on some key projects and making outstanding achievements. However, there had not been much improvement in the technology of production. The situation was certainly due to the faults of the economic system and economic policy, but it reflected ignorance about the effects of research and development on the technology of production. With a view to strengthening research on the technology of production, the Report Outline stresses a few issues that were ignored in the past. First, the purpose of research on the technology of industrial and agricultural production should be to promote the development of commodities with good quality and lower cost, as needed by the people. Second, attention should be paid to getting the relevant technologies well matched for the formation of a whole set and to getting the research, design and production well organized to keep the research results from remaining merely an exhibit. Third, economic benefits should be assured, and technology assessment and the analysis of economic results should be performed from the selection of research topics. All the three points made in the Report Outline are in connection with the weak points of technology research under a planned economy.

The Report Outline points out that regarding research work on industrial technologies, an overall arrangement and reasonable disposition should be made with respect to near-term and long-term needs as well as to basic research, applied research and development research. For the time being, special attention should be paid to development research, and strong efforts should be made to improve the quality and performance of basic components and materials as well aş to enhance the technological level of basic technologies. The selection of technologies should never blindly pursue what is brand new and on the cutting edge, but should proceed from the requirements of production and possibility to select appropriate advanced technologies. According to the actual circumstances in China, over a long period of time, the technical makeup of industrial and agricultural production will be a multilayered structure, with the coexistence of automation, mechanization, semimechanization and even manual labor. With a view to selecting technologies properly it is necessary to make a comprehensive comparison and analysis of the following factors: whether the choices can make good use of China's own resources, save on energy and the consumption of raw materials and enhance the level of technology and productivity; whether the choices expand employment and improve people's standard of living; whether the choices can help remold the imported and exported product mix, enhance the foreign exchange earning rate and save foreign exchange expenditures and whether the choices can reduce or avoid environment pollution and ecological damage.

3. It is necessary to strengthen the development and extension of technology in industrial enterprises. The Report Outline points out that the scientific and technological work in manufacturing enterprises is a weak link. The technological progress and renovation of several hundred thousand enterprises will in no way be able to rely on a small number of research institutes. Scientific and technical personnel as well as the workers and farmers working on the forefront of industrial and agricultural production must be encouraged and supported to actively carry out technology renovation and extension and to develop new technology, and their accomplishments should be rewarded.

The productive conditions and management level of the existing enterprises differ in a thousand ways from one another, their products are diverse in patterns and varieties and their demands for technology renovation differ from one another immensely. Only by carrying out technology renovation in manufacturing enterprises in an extensive way, adopting new technologies actively, improving the quality of products and speeding up the renewal of products can these enterprises achieve technology progress constantly, improve their competitiveness incessantly, adapt themselves to the changing market and rapidly increase their economic benefits.

The technology research of enterprises should be developed gradually, according to the actual circumstances and possibility, by giving full play to the role of the existing scientific and technical personnel of the enterprises and making good use of the internal material conditions of the enterprises. Every enterprise should give its own superiority full play and pay attention to cooperation with other enterprises, research institutes and higher education institutions, to absorb and purchase their technology results.

The scientific and technical forces of research institutes and higher education institutions should be encouraged and organized to actively support the technological research work of enterprises. Such research work may be carried out in the forms of accepting entrustment from industries, cooperative research and setting up various kinds of combined research and production entities, and so on. Some research institutes may be merged with manufacturing enterprises. The scientific and technical personnel should be encouraged to visit enterprises to offer assistance, doing concurrent jobs or to offer consulting service.

4. Basic research should be enabled to develop gradually on a stable basis. The Report Outline points out that basic research is an important part of scientific and technological research. The task of the research in natural science is to recognize the laws of Nature and to further master and utilize the laws of Nature. Economic development and technology innovation require a certain reserve of basic research. A country must have her own knowledge base of natural science, and some work can be undertaken only by the scientists of her own country. Some basic research plays an important role in the overall development of modern science. On the other hand, corresponding basic research should be carried out from the aspect of the enhancement of the level of higher education. Currently, it is necessary to emphasize strengthening basic research with relevance to national economic development and improving the quality and effectiveness of research.

Currently, expenditure for basic research in China is limited, accounting for only 5% of the total budget of civilian research. As the cycle of basic research is always longer, it is necessary to maintain its stability and succession. A basic research project cannot be started or cancelled hastily and carelessly. A long-term view is necessary to basic research. Any change made in terms of basic research should be made through peer review. In the coming decade, only a limited number of basic research projects will be carried out.

5. The learning, assimilation and absorption of foreign scientific and

technological achievements should be taken as an important path for the development of China's science and technology. The Report Outline points out that it is necessary to give a sober account of the gaps between China and the advanced level of the world. China should actively learn advanced science and technology abroad, with the aim of maintaining independence and keeping the initiative in her own hands. During a certain period of time, emphasis should be laid on learning, assimilation and absorption of advanced foreign technology appropriate for practice in China. Selection should be exercised among the technologies to be introduced, not just to pursue everything brand new, but to proceed from the economic requirements and technological basis of China.

NOTES

1. In 1988, Deng Xiaoping further emphasized the importance of science and technology and put forward the thesis "science and technology is the first force of production" (or "top productive force"). Gong Yuzhi's article "On Science and Technology Being the First Force of Production" has a detailed description of the background of Deng's thesis.

2. "Outline of the Report on the Guidelines for China's Scientific and Technological Development" was created by the China State Science and Technology Commission and submitted to the State Council on February 23, 1981. The report was approved and announced by the State Council later.

3

The Reform of the Management System of Science and Technology

After the general orientation of science and technology policy is fixed, it is essential to develop a new and effective mechanism to implement it. Usually, the development of science and technology needs a driving force from two different ends: technology push and demand pull. In order to strengthen technology push, it is necessary to stimulate the initiative of scientific and technological research institutions, as they are the chief producers and providers of new technology; and in order to promote demand pull, it is necessary to give an impetus to enterprises, as they are the chief users of technology results. With a view to making good use of both technology push and demand pull to quicken the pace of scientific and technological development, people must utilize economic leverage, market mechanisms and competition. That is why the original mechanism under a planned economy should be reformed and a mechanism suited to the new economic system should be developed. The reform of the science and technology management system was thus put on the government's agenda.

In 1985, the "Decision of the Central Committee of the Communist Party of China on the Reform of Science and Technology Management System" was announced. When the Decision was announced, the government leader pointed out that China's original science and technology management system was formed under specific historical conditions. Although it concentrated forces to solve some major scientific and technological issues, it was inadequate in meeting the demands of building the four modernizations by relying on science and technology. One of the major faults of the management system was the divorce of science and technology from production. There should be manifold cross channels between scientific research and production. However, the old management system blocked the channels for direct linkage between the two. Research institutions were under a vertical system in which they were responsible only to their senior departments, without any channels to lead to the society and serve production. Scientific research institutions had long failed to conform to the demands of production. Even research results needed by production had been difficult to apply and extend to production, because of either high cost or inability to outfit technologies into a whole set. The only way out was to carry out reform and build up a new science and technology system.

The Decision indicated that emphasis should be laid on how to fit the science and technology management system to the planned commodity economy in China. The role of commodity monetary relations, the law of value and economic leverage should be brought into play in promoting the development of science and technology.

The major points of the new science and technology management system were covered by the Decision.

1. The appropriation system[1] applied in research institutions should be reformed, and a diversified system of funding should be carried out in accordance with the nature of different categories of research activities. According to the Decision, the reform of the appropriation system does not imply a reduction in the science and technology appropriation in general. Rather, funding should be increased gradually, faster than the rate of growth of the regular financial income of the State. At the same time, funding sources should be broadened, and science and technology investment by various departments, enterprises and social groups should be encouraged.

2. The commercialization of technology results and the development of a technology market should be promoted in a way that conforms with the development of a socialist commodity economy.

The Decision pointed out that science and technology was the product of intelligent labor of mankind, and the value created by intelligent labor should be fully recognized and evaluated. With the development of science and technology, the role played by technology in creating the value of social commodities becomes ever more important. The technology market would be an important component of China's socialist commodity market.

According to the Decision, it was necessary, through the development of a technology market, to facilitate channels for the flow of technology results to production and to change the practice of transferring technology results by administrative command, without compensation. Such a change in the operational mechanism would favor linking up research institutions and production units by common interests as well as promotion of competition. This change would help translate the demand by production for scientific and technological assistance into research topics and to apply the research results in a timely manner to production.

3. It is necessary to readjust the organizational structure of the science and technology system, to encourage the combination of research, education and design institutions with production units and to strengthen the capability of industrial enterprises for technology assimilation and development.

According to the Decision, research institutions devoted to technology development under the Chinese Academy of Sciences, universities and various ministries and local governments should be encouraged to combine with enterprises in various forms, according to the principle of volunteerism and mutual benefits.

When fully relying on outside scientific technological forces, industrial enterprises should actively strengthen their own technology development capability.

The Decision also addressed other important issues, as follows:

1. The defense research institutions should establish a new system for the integration of military research with civilian research, to be oriented toward economic construction while ensuring the fulfillment of defense tasks and to accelerate the transfer of military technology results to civilian products.
2. The agricultural science and technology system should be reformed to favor the readjustment of rural economic structure and to promote the transformation of the rural economy into specialization, commercialization and modernization.
3. The in-depth disposition of scientific research should be arranged to ensure the follow-up strength of scientific, technological and economic development. Both basic and applied research were considered.
4. The scientific and technical personnel management system should be reformed to create a good environment for the emergence of thousands upon thousands of people of talent and for people to fully display their talents.

The Decision included comprehensive instructions about the reform of the science and technology management system. The core of the Decision was to indicate how to develop a new mechanism favorable to scientific and technological development in China. The most important aspects were the reform of the appropriation system, the development of a technology market that would serve the commercialization of technology results, the improvement of the technology assimilation and development capability of industrial enterprises and the reform of the scientific and technical personnel management system. A more detailed account of these aspects will be given in the following chapters.

NOTE

1. The appropriation system here stands for the original way of funding scientific and technological research in China. Under the central planning economic system in China, all the research funds were allocated by the government through the administrative system. All the institutes under central ministries or local government, as well as the Chinese Academy of Sciences, were funded by the government. For instance, each ministry had its research budget and was funded by the ministry of finance. The State Science and Technology Commission was in charge of the overall coordination of the science budget and was responsible for State major research projects directly.

4

The Reform of the Appropriation System

With a view to making technology push effective, the key is to give incentives to research institutions and to stimulate their initiative. The original appropriation system under the planned economy was ineffective in this regard. Under the original system, all operating funds of the research institutions came from State appropriation and all research tasks were assigned by senior government departments. This was a vertical system, and it was parallel to industrial production, which was under a different vertical system. The old appropriation system was suited to a planned economy, as all research tasks were under an overall plan mapped out by the planning department.

The system brought about a number of negative effects. For instance, as all research tasks were assigned by a senior government department, it was difficult for all these tasks to conform to the practical demands of production. How could tasks assigned by planners be guided by socio-economic demands? This process couldn't but bring about the divorce of research from production. On the other hand, since the distribution of operating funds was based on the size of each institution, including the number of its staff, without considering its research results, research institutions were responsible only to the senior department and ignored the production units to which they should be rendering their services. As all operating funds came from State appropriations, whenever the State was short of funds and failed to finance all research tasks, the completion of research projects was delayed and the research institutions could not run at full capacity. Research staff couldn't give their potentiality full play, and it was difficult for many small and medium-sized enterprises to get technical force needed to solve their technical problems. That was a big waste of precious human resources.

Furthermore, the old appropriation system applied a single form of management to different categories of research. Basic and applied research and development research were all supported by State appropriation without competition or strict selection. No incentive could be given to research faculty, since the research faculty had no need to try their best to win support for their specific research projects. This system did not encourage scientists to improve their working efficiency and enhance their research level. Above all, the old appropriation system was not a productive one and did not favor the integration of research with production; it made the transformation of scientific and technological force into a productive force difficult. The only option was to replace it by a new effective funding mechanism.

The reform started by breaking up the single appropriation method and diversifying funding resources. Under the new system, research funding was mainly divided into three categories.

First, the funding of development research and applied research that might bring about near-term practical value was separated from basic research and from applied research in general. If work by researchers might enable the users of their technological results to get near-term benefits, they could push those results onto the market as commodities and produce revenue. Cutting off such research institutions from State appropriations meant that they should "earn their living" by their own intelligent labor; the more and better the technology results they turned out, the more benefits they might get. In other words, they should be commodity producers instead of government functionaries. Certain pressures would be imposed and certain incentives would be given to these institutions and their faculties.

Second, as basic research and applied research in general do not create any short-term benefits, such research should have the government and society as its "patron." However, should the institutions devoted to such research just "live" on State appropriation as they did before, there would be no pressure or incentive for them. The solution was to let them be funded by various foundations. The process of application, peer review and approval, as well as the follow-up examination of research progress, not only imposed certain pressures on these institutions but also brought about competition among scientists. Scientific life under the new system would favor scientists who were more aggressive in making advances in their research.

The third category is research devoted to social welfare, which should be supported by the government and the society.

With the old appropriation system broken up, the government correspondingly changed its original way of assigning major projects listed in the State plan according to administrative command. As stipulated by the State Council, all major science and technology projects should be sent out for bid. All institutions joining the bidding should be guaranteed equal rights and equal opportunity; and in turn, all institutions who win the bidding should ensure that the performance and indicators of their research results meet the requirements of the State plan. To facilitate implementation of the new system, corresponding policies were formulated, such as a new accounting system suited to the new appropriation system and the establishment of science and technology credit funds in China People's Bank to help research institutions get working funds for their projects.

With various conditions ready, the new appropriation system was carried out as planned. By 1990 and 1991, most technology development research institutions had had their operating funds cut, as required by the new appropriation system. These institutions had made great progress in shifting their research to serving industry directly and had achieved considerable growth in revenue through technology contracts and other technological development projects with industry. For instance, in 1987, the research institutes under 57 ministries had a net income of 594 million yuan, accounting for 37.85% of the total operating funds allocated by the State, an increase of 23.8% over 1986 (450 million yuan). In 1988, the research institutions devoted to development research all over the country created a total gross revenue of 38.3 billion yuan, for the first time exceeding the operating funds provided by the State. The scope of commercialized technology research activities had already exceeded the aggregate quantity of those supported by traditional administrative funds.

The transformation in funding brought about a series of changes in the scientific and technological community. First of all, a market and competition mechanism was introduced into scientific research and imbued the research community with new vigor. Previously, scientists stayed in their "ivory tower" without any pressure or incentive to enter social and economic life. Now, it was quite different. Scientists and engineers turned much more aggressive, with a competitive mentality. They began to open their eyes to the market.

Second, the operational system of research institutions changed. Most of the development research institutions changed from the previous closed loop research system into a new open research and operation system. A number of such institutions set up new departments exclusively dealing with market development. In the meantime, the old mentality under the original appropriation system was changed, too. For instance, the mentality of dependence on senior government departments was watered down, and ties with industry became closer.

Third, the reform of the appropriation system was closely related to the development of a technology market. Without a matured technology market, it would be difficult for research institutions to find buyers for their technology results.

Fourth, the reform of the appropriation system helped to promote the mobility of scientific and technical personnel and caused some research institutions and higher education institutions to move into production and economic circles directly.

Fifth, the reform of the appropriation system promoted the emergence of a new technology investment system. As science and technology were oriented toward the economy directly, the previous investment model had to change. Under the original investment model, State appropriation was the only source of financing for technology development. Now, a brand new technology development investment model was developed. It was comprised of diversified and multilayered investment channels, including investment from central level and local level enterprises and financial institutions. Bank credits played a role as catalyst in the process of integrating science and technology with production. The science and technology debts of the banks increased from 10 million yuan in 1984 to 1,870 million yuan in 1988.

5

The Development of a Technology Market

The purpose of developing a technology market is to promote and facilitate the integration of scientific and technological research with production. Technology can be a commodity in the shape of knowledge; but without a market, technology cannot become a real commodity. It is necessary to provide a marketplace to make technology change hands between research institutions as sellers and enterprises as users. Only by bringing both parties together can the value of technology be evaluated and realized.

A process for developing a technology market in China had existed. As early as 1978, technology trade actually started, a prelude to the development of a technology market. Of course, during that time, technology trade was still in a primitive form. Usually, it was a research institute that entered into cooperation with an enterprise on a voluntary basis to tackle a certain difficult technological issue or offer technology services and consulting for compensation from the enterprise. That was the start of compensated transfer of technology. In 1982, when science and technology began to be oriented toward economic construction, a coordinated development of science and technology with the economy and society speeded up the formation of a technology market. The year 1985 was important in the process of developing a technology market. In January of that year, the State Council issued the "Provisional Regulations on Technology Transfer"; and in March of the same year, in the "Deci-

Table 1
The Growth of the Amount of Technology Trade Between 1983 and 1989 (in 10,000 Yuan)

Year	Volume of Technology Trade
1983	5,000
1984	72,000
1985	230,000
1986	206,000
1987	335,000
1988	724,881
1989	814,639

Source: "Guide to China's Science and Technology Policy."

sion of the Central Committee of CPC on the Reform of Science and Technology Management System," it was further pointed out, "A technology market is an important component of the socialist commodity market of our country." All these events consolidated the status and role of the technology market. A major breakthrough had been made in respect to the principal issues of the commercialization of technology results; the technology market had rapid growth thereafter (see Table 1).[1]

With the growth of technology trade, the extension and application rate of technology results had gone up correspondingly. The growth of technology trade needs a broad technology business network to bring sellers and buyers together and to facilitate information exchange. As well as the existing research institutions at various levels, it needed the development of a large number of institutions devoted to technology extension as well as a large number of intermediary institutions to help promote technology trade. In a few years, a large number of institutions devoted to technology development, technology trade and technology intermediary services were developed.[2]

By setting up all these institutions, as listed in Table 2, a multilayered, multichannel and diversified system was established. What was now needed was to bring many of these organizations together to form a network with a broader or nationwide scope. In this regard, the Nationwide Science, Technology and Personnel Development and Exchange Coordination Network was established. The network was sponsored by the China Technology Market Management and Promotion Center and co-sponsored by a number of relevant institutions in various provinces and municipalities. Under the direction of the nationwide coordination network, the Technology Market Network System was established in the northwest, southwest, central south, east, north and northeast of China

Table 2
Development of New Institutions Devoted to Technology Development and Trade

Various Kinds of Technology Development and Trade Institutions above Prefecture and County Level	5,000
Various Kinds of Research and Production (Combined Entities)	10,000
Various Non-governmental and Individual Research and Technology Development Institutions	10,000
Professional Societies Established by Farmers	72,000

Source: "Guide to China's Science and Technology Policy."

successively. The Chinese Academy of Sciences, the Ministry of Nuclear Industry and the State Administration of Building Materials also established their own technology market networks. In the meantime, the science and technology development centers under various central ministries had jointly established the United Technology Market Development Group of China which, being a powerful technology development group and an organization serving information exchange and technology renovation in various parts of the country, was a multidisciplinary, trans-industrial-departments–combined entity with relaxed connections.

With the rapid growth of technology trade and the formation of a broad network, the forms of technology trade had been diversified and turned out to be more flexible. The following practices were effectively adopted for the promotion of technology trade:

1. To sponsor various specialized technology trade conferences to carry out specialized technology trade.
2. To hold information tea parties for the promotion of information exchange.
3. To contract for a complete set of technology services.
4. To set up a standing technology fair, and to organize the full-time or part-time scientific and technical personnel to carry out activities of technology development and technology service.
5. To start up a science and technology demonstration factory to strengthen the capabilities of intermediate experiment, to accelerate the commercialization of technology results.
6. To start up a scientific and technical personnel development tank to promote the reasonable flow of scientific and technical personnel.
7. To set up science and technology development funds to make use of economic leverage to promote regional economic development by relying on technological progress.

With the development of a technology market, a buyers' market had been formed. It was owing to the further development of economic reform that more enterprises were attracted to the technology market. The extensive participation of enterprises as buyers had changed the situation of the technology market. After 1986, only 10% of technology trade was spot transactions (the purchase of ready-made technology results), and more than 80% of business transactions were sales on customized orders (the purchase of the technology tailored to the demands of specific buyers).

Research and development institutions and industrial enterprises were the biggest sellers of technology in the technology market. The proportion of research and development institutions as technology sellers had further increased. These institutions had been the beneficiaries of the technology market. They had expanded the application of their technology results, increased their revenue and strengthened their self-development capability through the technology market as an important medium.

With the development of the technology market, the varieties of technology commodities were also diversified. More projects that were listed in the plans of State and local governments entered into the market for public bidding. A number of engineering projects also entered the market to find the best contractors.

A number of projects listed in the plan for technology importation were now turning to the technology market for bidding, to tap the latent domestic technological power.

More technology commodities appropriate for use by small and medium-sized as well as township enterprises entered the technology market. Many research institutions and higher education institutions had exclusively developed appropriate technologies for small and medium-sized enterprises and township industries. These were well received on the technology market. A number of technology commodities with respect to light industry, the food industry, and the building materials and petrochemical industry, which were closely related to the improvement of people's living standards with higher economic benefits, were in great demand on the technology market.

However, the level of technology commodities still required upgrading. For instance, technology service contracts accounted for a higher percentage of work, while research and development contracts and technology transfer contracts accounted for a lower percentage (see Table 3).

The development of a technology market brought about a number of changes. First, there was a change in mentality. Previously, under the planned economy, no value had been credited to the knowledge product. Now it was different. "Technology has become a commodity, can be

Table 3
Composition of Different Kinds of Technology Contracts

Content of Contract	Number of Contracts	Percentage	Amount of Payment (in 10,000 yuan)	Percentage
Technology Service Contract	90,811	75.31%	179,761	59.03%
Research and Development Contract	14,155	11.74%	83,371	27.38%
Technology Transfer Contract	13,441	11.15%	39,308	12.91%
Technology Export Contract	2,170	1.80%	2,066	0.68%

Source: "Guide to China's Science and Technology Policy."

exchanged and can be bought and sold." The new mentality had become popular very quickly with the development of the technology market.

The development of a technology market helped to promote the concept that science and technology should be oriented toward the economy and the market. It was no longer a concept on paper only. It penetrated people's daily work. Previously, under the planned economy, aside from the fulfillment of the tasks assigned by the senior department according to overall planning, scientists did not choose their research topics according to the demands of practical production, but chased after "overseas tendency" and picked their research topics from foreign scientific documents. In the process of technology development, the selection of topics through the extension and application of technology results, scientists and engineers gradually turned to the market and social demand for direction. Scientists and engineers switched to the main battlefield of economic construction, contributing to the study and solution of the issues in respect to technology renovation and innovation in production and bringing science and technology into close integration with economic construction.

The scientific and technological research institutions had been the main beneficiaries of the development of the technology market. With scientific and technical personnel integrating their work with economic construction and taking market demand as their direction, research institutions were trying in every way to combine their contribution to society with their own economic benefits. When research institutions entered the market, they gained experience in business operation, of which they had long been ignorant in the "ivory tower." Most institutions set up their own project development departments in charge of cross linkages with enterprises and other clients. Instead of focusing on technology transfer, a number of research institutions had gone further, to set up combined research and production entities with enterprises. A number of research institutions merged either with large industries (to become the R&D departments of these large enterprises) or with some small industries (to own these small enterprises and use their production facilities to develop new technology intensive products). It gave rise to quite a number of knowledge-based industries. The reform of the appropriation system was aimed strongly motivating research institutions to strive for a new way out. Now, the development of a technology market provided them with a brand new world that let them fully display their ability.

The reform of the science and technology management system was quite successful in strengthening technology push, but it turned out to be less successful in strengthening the other end, demand pull. The small and township industries played a very active part in the technology market and helped to invigorate the market. The large and medium-sized enterprises had taken part in the technology trade to some extent and taken a larger share of technology contracts. However, relative to the total number and size of large and medium-sized enterprises, their participation in and contribution to the technology market were limited, not only limiting the business volume of the technology market, but also hindering the upgrading of technology trade. Actually, large and medium-sized enterprises possessed great potential for developing technology trade. Insofar as these industries possessed superiority in technology and equipment, it was much easier for them to accept, assimilate and develop new technology at a higher level. And after absorbing newer and higher technology to promote their own technological progress, they could organize their technological forces to make new innovations and to develop new technology and new products and let these flow back to the technology market. It would follow a normal cycle. However, as the reform of the management system of large and medium-sized enterprises lagged behind and for some other reasons in terms of external economic environments, the demand for technological progress was not strong inside large and medium-sized enterprises. It was impossible for the technology market alone to attract these industries to take an active part in technology trade.

Small and township industries were also beneficiaries of the development of the technology market, as they took advantage of technology trade and got the necessary technologies they badly needed. These industries, especially the township industries, were short of both technology and technical people. The ratio between engineers and workers in township industries was nine per 10,000. The introduction of appropriate technologies into township industries made it possible for them to upgrade their technology and product level. Quite a number of township industries were able to get access to the export market after their product level was upgraded by technology purchased from the technology market.

Research institutions and their scientists and engineers not only found opportunities in small industries but also had new fields to exploit. The vast territory of the rural areas needed new technology badly. The Spark Program sponsored by the State Science and Technology Commission was aimed at the development of appropriate technologies to meet the demands of farmers and township industries. On the other hand, the development of high and new technology also opened a new field for the scientists and engineers.[3]

The Torch Program sponsored by the State Science and Technology Commission was aimed at high and new technology development as well as the development of a series of rising industries. The emergence of new fields helped to expand the technology market and allowed it to exploit the new technology trade in greater depth.

As the technology market has developed for more than a decade, the current situation is quite different. The development of the technology market started with the purpose of overcoming the divorce of scientific research from production as well as pushing scientific research into the main battlefield of economic construction. Now, the situation of research institutions is quite different. They not only have improved their integration with production and take an active part in economic construction but also directly participate in industries, either as technology shareholders or as research and development arms of the industries. Furthermore, they are starting or running their own technology businesses and have become insiders rather than outsiders in industrial development. How far the research institutions have come!

NOTES

1. The figures for Table 1 are from "Guide to China's Science and Technology Policy" (White Paper on Science and Technology), which is an official publication, compiled by the State Science and Technology Commission and published by the Literature of Science and Technology Press, Beijing, China. The publication provides information on scientific and technological development in China and various official documents related to science and technology policy. It was first published in 1986. Up to now, seven issues have been published.

2. After the realization of the output related responsibility system in the rural areas, farmers had the initiative for farming, but a great deal of farming still followed the traditional way. When local scientific institutions started to promote new agricultural technology, a large number of farmers were enthusiastic in learning and adopting new farming methods. A number of local professional societies for disseminating new agricultural technology were developed.

3. The term "high and new technology" covers not only high-tech, but also some technologies that do not exactly belong to the category of high-tech but are new to China.

6

Strengthening the Capability of Technology Development and Assimilation of Industrial Enterprises

The internal demand from industrial enterprises for technological progress is an important driving force in the promotion of industrial technology. The reform of the appropriation system helped intensify technology push, and the development of the technology market provided enterprises with a good environment to source and secure new technology. However, the internal demand is dependent on both the external economic environment and the internal operational management system.

After about three decades of industrial development, China had built up an initial industrial system of a certain scientific and technical force. By 1988, there were 10,738 large and medium-sized enterprises in China with a total annual output value of 739.83 billion yuan, accounting for 40.6% of the total industrial output value of the country, and the total amount of profits and taxes turned out by all these enterprises was 148.68 billion yuan, accounting for 83.77% of the total annual industrial profits and taxes of the country. Together these industries played a decisive role in China's national economy. The promotion of the technological progress of these industries was also a critical issue for both economic reform and scientific and technological reform. Unlike small and township industries, large and medium-sized enterprises did have certain scientific and technological force.

By the end of 1988, there were 1,585,000 engineers and technicians in

Table 4
Technology Purchasing and Selling by Large and Medium-Sized Enterprises in 1987

	Number (and %) of Contracts	Total Amount of Revenue and Payment (in 10,000 yuan and %)
Technology Selling	6,410	31,203
Proportion to Total Number of Contracts and Total Amount of Revenue	6.6%	13.3%
Technology Purchasing	30,975	101,936
Proportion to Total Number of Contracts and Total Amount of Payment	32.1%	43.3%

Source: "Guide to China's Science and Technology Policy."

large and medium-sized enterprises, accounting for 5.99% of total workers and staff. This included 797,000 engineers and technicians who were engaged in technology development, and among them 269,000 were scientists and engineers. The technology development expenditure was 11.6 billion yuan in total for all these large and medium-sized enterprises, accounting for 1.56% of the annual sales of all these industries. In R&D projects, there was more product development than technology development. The sales of new products accounted for 7.64% of the total sales. All the figures boiled down to one point: It was essential to further improve the capability of developing and assimilating new technology of large and medium-sized enterprises. However, the data on technology purchasing on the technology market by these enterprises showed limited progress.

It can be seen in Table 4 that the proportion of technology purchasing of large and medium-sized enterprises was limited in relation to their proportion in output value. The total spending of all these enterprises in purchasing new technology was less than 0.14% of their total annual output value. The relative passivity of the large and medium-sized enterprises in the technology market stemmed from weak internal demand in these enterprises for technology progress. The weaker internal demand derived from both the external environment and the internal management of these enterprises.

The internal demand of industry for technological progress needs certain external pressure. A sound market mechanism and competition would help to promote an industrial effort for technological progress. In

Table 5
Sales of State-Owned Enterprises

Year	Sell Well	Ordinary Selling	Unsalable
1986	28.0%	52.1%	19.9%
1987	53.8%	41.2%	5.0%
1988	51.2%	40.5%	8.0%

Source: "Guide to China's Science and Technology Policy."

China, under the original central planning economic system, instead of a commodity economy, what had been exercised was the "product economy." What the industry turned out was not commodities but only product. Industries accepted the assignment of production targets from the senior planning department and concentrated their efforts on fulfilling quantitative targets. Therefore, a "product economy" is also a "quantitative economy." Because the products produced by industries were completely transferred to government distribution channels, industries had no need to worry whether there was a market for their products. Production without competition, of course, didn't bring about any serious concern for the constant improvement of quality, as the quality standard issued by the government was not dynamic, but static. Likewise, production without competition meant no incentive for industries to carry out innovations for the reduction of energy and material consumption and the constant enhancement of productivity, to bring down costs. Without competition, industries felt no need to strive for the development of new products and hence the renewal of technology. That is one reason that China's industries under a planned economy had kept their outmoded technology, techniques and equipment for so many years.

The economic reform had been trying to change such a situation and to bring industry out of an abnormal cycle. However, during the 1980s, though the economic reform had partly brought the industries to take part in a market system, the market in China was still immature. It was still under a certain influence of the original planned economy. For instance, the planned economy had always been a "shortage economy," as demands always exceeded supply, and imposed no serious pressure on the industries.

Table 5 shows that under the shortage economy, selling was in a good state, especially as, according to some sampling investigation, the goods with competition accounted for only 44%, while the goods with exclusivity accounted for 47.6%. Of producers, 8.2% felt that they had no competitors, and 77% of producers who felt in competition found that

77% of the competition came from the same kind of goods; competition from substituted goods accounted for only 22.4%. The circumstances offered little to stimulate the desire for technological improvement.

Aside from an external economic environment, the reform of internal production organizational structure may also help to promote initiatives for technology renovation and technological progress. By the late 1980s, 90% of industrial enterprises all over the country had implemented a contracted responsibility system. Under this system, the management team of an enterprise was required to sign a contract with its senior authority taking responsibility for the operation of the enterprise; the specific responsibilities, power and benefits of the management team were all listed in the contract. The contracted responsibility system served as a tool to separate operation from ownership. Previously, as each enterprise was owned by the State, its operation was under administrative control of its senior authority, and the management team had only limited power and hence limited responsibility. By separating operation from ownership, it was expected that the management team would have full power and hence full responsibility for its operation. The interests of both the management team and the workers and staff of the enterprise were to be bound together with the success of operation. Henceforth, it would be possible to establish an operational mechanism with a stronger capability for self-development and self-restriction.

However, during the process of implementation, the contracted responsibility system did not function as well as expected. The separation of operation from ownership and hence the divorce of enterprise from administration had been a major point for the contracted system, but had not been carried out thoroughly. The relationship, in terms of benefits and restrictions between enterprise and administration, was not strictly defined. As property rights were still owned by the State, the enterprise itself was not in a position to take direct responsibility for the assets of the enterprise. All these factors allowed the contracted system control over only short-term behavior. That is to say, what the contracted system could cover was only the near-term benefits of enterprises and their management teams, and what the contracted system could stimulate was only initiatives for near-term efforts. The system did little to encourage enterprises to strive for long-term prospects, including the development of programs for technological progress that offered no benefits in the near term.

Notwithstanding the further growth of State-owned large and medium-sized industries in quantity, the situation in connection with the quality and economic benefits of these industries deteriorated further.

Table 6[1] gives a recent account of the strength of these industries that shows that they have grown in quantity by a big margin and possess a virtually decisive portion of national industrial power. The problem is

Table 6
State-Owned Large and Medium-Sized Industries

Total Number of Large and Medium-Sized Industries	23,007 (Including 215 Super-Large Industries, 6,201 Large Industries and 16,591 Medium-Sized Industries)	Proportion in the Total Number of Industries All Over the Country	0.3%
Total Number of Workers and Staff	38.91 million	Proportion in the Total Number of Industrial Workers and Staff	26.4%
Total Assets	5,220.53 billion yuan	Proportion in Total Industrial Assets	59.1%
Total Sales	3,084.71 billion yuan	Proportion in Total Industrial Sales	39.9%
Tax Payment	286.05 billion yuan	Proportion in Total Industrial Taxes	61.6%
Total Amount of Profits	129.13 billion yuan	Proportion in Total Industrial Profits	79.0%

Source: China's Third Nationwide General Industrial Survey by the State Statistical Bureau in 1996.

that the growth is only quantitative and only derived from the increase of input and the expansion of size. Without substantial improvement at the technological and managerial levels, such growth cannot lead to the drastic cutting of material and energy consumption and the outstanding enhancement of the quality of products, and thus is unable to bring about any conspicuous improvement in economic benefits. As a matter of fact, with the further development of market mechanisms and the intensification of competition, the monopoly advantages of these large State-owned industries are decreasing and their economic benefits have dropped seriously. This situation must bring about a serious setback in China's efforts to build up a modernized industry and to rejuvenate the national industry by relying on technological progress. The decisive preponderance of these industries in the national economy makes the problem even more serious. The only way out is to further China's economic reform with a view to solving the problems derived from the industrial system itself and propelling these industries to shift onto a new track to strive for qualitative growth. They need to improve their economic benefits by relying on technological progress, which may help to lower their material and energy consumption and raise the quality of their products. Only by fundamental change in the industrial system can the internal demands of technological progress be stimulated and create a strong momentum of demand pull in these State-owned industries. In consideration of the proportion of these industries in the national economy, the change will certainly be tremendous, not only in the prosperity of the national economy but also in the development of industrial technology.

NOTE

1. The big difference between the figures provided in Table 6 and those in the beginning of this chapter is due to a couple of reasons: growth in the number of large and medium-sized enterprises between the 1980s and the 1990s and a high rate of inflation over the decade.

7

Reform of the Scientific and Technical Personnel Management System

The reform of the management system of scientific and technical personnel[1] is an important part of scientific and technological reform. Scientists and engineers are developers of science and technology. Therefore, when Deng Xiaoping first announced the grand goal of four modernizations and emphasized the key role of the modernization of science and technology, he at once reevaluated the role of scientists and engineers and gave an important account of the status of intellectuals in China and called for people to respect knowledge and people of talent. This played an important part in restoring the reputation and improving the social and political status of intellectuals. It gave great encouragement to scientists and engineers and helped them contribute to the economic construction and to the building of modernization with enthusiasm and initiative.

However, scientists and engineers had been tied up by the central planning economic system, a very difficult environment in which give their ability and potential full play. Under a planned economy, every scientist or engineer was fixed to a position either in a government agency or enterprise or in a research institute and higher education institution. People always kept a lifelong position at one institution, without chance to move to some other organization that might give them more opportunity to display their ability.

China had had a rapid growth in the number of scientific and technical personnel after the founding of the People's Republic. The size of the scientific and technical force when the People's Republic was first founded was insignificant, but on the eve of the Cultural Revolution, China had already had a relatively stronger scientific and technical force. The Cultural Revolution cut off the continuous process of the training of scientists and engineers for ten years; but after the Cultural Revolution, progress in science and engineering education resumed at full speed; during the period between 1978 and 1984, the average annual increase in scientific and technical personnel was 6.5%. By 1987, the total number of scientific and technical personnel came to 8,894,000, including 4,172,000 scientists and engineers.

The problem lay in the unreasonable distribution of these scientific and technical personnel. China's scientific and technological force was growing under the condition of emphasis on industrial development, especially on the development of heavy industry. It had been inevitable for the country to concentrate a relatively stronger scientific and technological force on the development of heavy industry and the defense industry for a time. However, because of the rigidity of the personnel management system, as well as the restriction derived from the original specialty of scientists and engineers, the scientific and technical force could not be redistributed or readjusted according to the change of focus of national economic construction. Such rigidity inevitably led to an unreasonable distribution of the scientific and technological force and thus made it impossible to meet the requirements of the development of national economic construction. Also because of the rigidity of the personnel management system, more scientific and technical personnel were concentrated in higher education institutions and research institutes and fewer were working with enterprises. According to statistics, the number of scientific and technical personnel working with enterprises was less than half the total number of scientific and technical personnel in the country in 1986; and quite a few key industries, such as the energy, transportation, telecommunications and material industries, suffered from the shortage of scientific and technical personnel. In recent years, a large number of township industries were developed. In these newly developed industries, there were only nine technical people per 10 thousand employees on average in the early 1980s. The focus of the reform of the scientific and technical management system was to relax the rigid control in the management of scientific and technical personnel, to promote the reasonable flow of scientists and engineers and to encourage them to give full play to their ability, to make full use of the existing scientific and technological force in China.

The mobility of scientific and technical personnel improved through the relaxation of the personnel management system. Movement in scientific and technical personnel in three years amounted to 280,300 people. That was unimaginable in the past. The transfer of full-time faculty members wasn't unlimited, but it was easier for scientists and engineers to take part-time jobs in their spare time or to take leave for a period of time. Previously, scientists and engineers were not allowed to have second jobs. Permission for scientists and engineers to have second jobs or concurrent jobs was also part of the personnel management system reform. The government issued special documents with specific stipulations regarding permission and encouragement for scientific and technical personnel to do concurrent jobs or to have second jobs. According to a survey conducted in 1989, 26% of scientific and technical personnel had concurrent jobs, and most of them were helping small and township enterprises. In 1988 and 1989, there were 470,000 scientific and technical personnel who came out from research institutes, higher education institutions and government agencies to the countryside, factories and even mountainous areas to offer technology services or to carry out technology contracts.

The development of economic reform required the further reform of the personnel management system. In 1988, a decision was made by the State Science and Technology Commission on the mobilization and organization of scientific and technological force to serve the implementation of economic development strategy in coastline areas. The decision pointed out that in the contemporary world, for the purpose of promoting modernization, it had been a global tendency to propel economic development by relying on science and technology and to take part in international competition and commodity exchange. China followed the world tendency to accelerate scientific and technological progress in the coastline areas, to help people in the areas integrate science and technology with the economy and to further the input of science and technology into the economy to promote the rapid development of coastline areas. The decision demanded that the research institutions and scientific and technical personnel be provided with more flexibility and encouraged large numbers of scientific and technical personnel to move to the coastline areas to run the local township industries by contract, or to take the lead in operating township industries or even to run township industries by leasing. Under the direction of the decision, a number of research institutes developed their subsidiaries in the coastline areas. All of these measures furthered the mobility of the scientific and technical force.

The mobility of the scientific and technical force could also be improved by integrating research institutions with enterprises. In order to promote the growth of science and technology in the economy, quite a number of research institutions had either merged with enterprises or set up combined science and production entities with enterprises. There were quite a few different forms for such integration (see Table 7).

Table 7
Different Forms of Integration of Research Institutions (above County Level) with Enterprises

	Number of Institutes	Local Institutes	Institutes Under Central Ministries	Institutes Under Chinese Academy of Sciences
Total	2,469	1,678	712	79
Merged into Large and Medium-Sized Enterprises	259	157	93	9
Taking Lead in Operating or Running Small Enterprises by Contracts	182	130	49	3
To Be Technology Development Center of Various Industrial Sectors	293	172	109	12
Combining with Designing Institutes or Enterprises as Engineering or Technology Outfit Companies	154	80	70	4
Combining with Enterprises for Technology Development	1,060	764	259	37
Other	521	375	132	14

Source: "Guide to China's Science and Technology Policy."

In the 1990s, with the development of the market economy, a large number of non-governmental science-and-technology-oriented businesses and new high technology industries had rapid growth. Under these circumstances, a large number of scientists and engineers were spun off from their parent institutions to join these new businesses. The mobility rate for scientific and technical personnel was higher than ever before. With the further reform of the scientific and technical personnel management system, scientific and technical personnel intermingled with the market, and science and technology were further integrated with the economy. In the 1990s, the total number of scientific and technical personnel increased from 23.14 million in 1989 to 31.28 million in 1994, a net increase of 8.14 million and an average annual increase of 6.944%.

However, the distribution of scientific and technical personnel remains a serious problem. As was discussed earlier, the original personnel management system under the central planning economy failed to match the disposition of scientific and technical human resources to the demands of economic development. Although in 1990 the scientists working with industrial enterprises accounted for 42% of the total scientific and technical personnel, the junior technical people made up 15%. During the period of the eighth five year plan (1991–1995), the situation did not improve, but worsened. The percentage fell to 39.7% by 1994, while those working with various government departments increased by a big margin. In 1989, the total number of scientific and technical personnel working with government departments increased to 1.51 million, an increase of 2.28 times in nine years and an average annual increase of 25%. By 1994, the total number of scientific and technical personnel working with government departments increased to 4.41 million, an increase of 2.91 times in five years and an average annual increase of 38%. The percentage of those who worked with government departments accounted for 14.15% of the total.

The disposition of the scientific and technical human resources in different industrial sectors was also disproportionate. According to the statistics in 1989, those working with primary industry accounted for 4.3% of the total scientific and technical personnel working with industry, those working with second industry accounted for 23.2% of the total and those working with tertiary industry accounted for 72.5% of the total. The disproportionate disposition of scientific and technical human resources was nearly the same during the period of the eighth five year plan.

Such problems could be solved only by improving the mobility of scientific and technical human resources and further developing the market economy. This would help to rectify the disproportionate distribution of scientific and technical personnel.

The reform of the management system of science and technology provided a dynamic mechanism for the integration of science and technology with production as well as for the science and technology to become part of the economy. However, what the mechanism provided was mainly the technology push aspect, whether it was effective or not, in which field it was effective, or to which extent it was effective remains to be determined by the intensity of demand pull. Scrutiny of the practice of science and technology policy in different fields will be significant.

NOTE

1. "Scientific and technical personnel" refers to the people who are involved in scientific and technical activity and who have diplomas at or above the level of polytechnic schools or else have acquired a professional status through their work experience.

8

China's Opening and Technology Acquisition Policy

The introduction and assimilation of foreign advanced technology is an important part of China's opening policy. China's opening policy is not a static but a dynamic one, with incessant development. The development of the opening policy is closely related to the evolution of economic reform. China's economic reform, since its first announcement in 1978, has evolved in three stages: a planned economy with market adjustment as a supplement; the implementation of a planned commodity economy and the development of a socialist market economy. During the process of evolution, the role of the market economy was expanded. During each stage of evolution, both the scope and the depth of China's opening policy were broadened and furthered.

The change in mentality was most important for the development of opening policy. During the later stage of the Cultural Revolution, there was the "Fengqing Ship Event."[1] The leftist group used the building of the Fengqing ship to stir up large-scale criticism over the "comprador philosophy" or "crawlism." What the leftists criticized was a "slavish mentality towards all things foreign" that advocated that in order to build up China's own ocean transportation system quickly, it was necessary to charter or buy foreign ships during a transition to building China's own ships. As a matter of fact, it had been and is still conventional in the world for a developing country or area to start its own ocean transportation business. Hong Kong had achieved good success in this field. However, according to the leftist view, such behavior was running counter to the principle of relying on one's own force. While the leftist line prevailed, the introduction of anything to do with foreign experiences or technology was considered the behavior of a slave to foreign things. Only by relying on China's own force could the country be constructed along a correct track. Should the absurd interpretation of relying on one's own force remain a principle, the door of China would not open. During the late 1970s, when the Cultural Revolution was over, an argument carried out on a large scale criticized the erroneous interpretation of relying on one's own force. It publicized the concept that China would only develop at a faster speed if she could adopt an opening policy to absorb advanced technology, management experience and capital from abroad. Deng Xiaoping indicated: "If socialism wants to win the comparative superiority over capitalism, it should boldly absorb, or take as reference, all the results of civilization created by mankind as well as absorb, or take as reference, all the advanced operational and management methods of various countries including capitalist countries of the contemporary world which reflect the law of modern socialized production." It was actually a campaign of enlightenment that was aimed at the change of mentality and paved the way for the implementation of the opening policy.

The change in mentality has been a long and incremental process. In the 1980s, China-foreign joint ventures had been gradually growing into an important form of technology importation. In the process of developing China-foreign joint ventures, people asked if it was correct to accept foreign investment. The acceptance of foreign investment meant the absorption of capitalist economic element into socialist China. Deng Xiaoping's answer was: "The absorption of foreign capital, foreign technology, even allowing the foreign countries to set up their factories in China may serve as a supplement to the socialist force of production we are developing. Of course, it may bring in some corrupted elements of capitalism, but so far as we are aware, we don't need to worry about it."

Furthering the opening policy, of course, was required for the economic reform. But without a change of mentality, it would be difficult to realize. However, mental liberalization only serves as a vanguard for the further development of opening up. Its practice still needs a number of conditions, including the establishment of a sound legislative environment with transparency, such as the formulation of patent law, joint venture law, regulations on foreign exchange control and the like; a good infrastructure, such as the convenience of transportation, telecommunications and information networking; and the improvement of the investment environment, such as the working efficiency of the government departments, the simplification of procedures for the approval of investment projects by authority, banking service, market study and consulting service.

The process of developing an investment environment has been a prolonged one. In Beijing, the first China-foreign joint venture enterprise was approved in October, 1979. By 1983, only 12 China-foreign joint venture enterprises had been approved in five years. The negotiation on the establishment of Beijing Jeep joint venture was a "marathon" race. For various reasons, the negotiation had taken 65 rounds, and four years to close the deal. Of course, it doesn't need to take that long to make a deal today. The procedure of approval for the establishment of businesses with foreign investment has been much improved.

Without the change of mentality and the progress of an opening policy as well as an improvement in the investment environment, the importation of technology could be only carried out on a small scale. However, expanding the importation of technology is different from commodity importation in general; it requires that the capability of technology assimilation, absorption and innovation be improved. Technology importation in China has undergone a process of evolution in the past two decades. Technology importation is a concept with broad content. In a narrow sense, technology importation, as transfer of technology from one country to another, usually covers the know-how or the purchase or licensing of patented technology. It includes the transfer of software and, in many cases, the purchase of hardware (such as production equipment and facilities). In order to allow the transferee to put the technology into production and acquaint the technical staff with the operation of equipment and facilities, technical training is also a part of the deal in technology importation. Usually, a number of provisions, including after-sale service, are included in contracts of technology importation. As for technology transfer between developed countries and developing countries, the procurement of more hardware is included in the package, as the latter usually cannot develop a whole production line with just the core technology introduced from abroad.

Technology importation is expensive and the payoff takes time, especially as the purchase needs to be paid for in hard currency and the hard currency reserve of a developing country is limited. Therefore, in many cases, joint ventures are a good way to make importation of technology more feasible. By the arrangement of a joint venture, the transfer of technology can be rewarded with a technology share in the joint venture firm, and the hardware transferred can be taken as an investment by the transferor. The superiority of a joint venture is that by setting up a joint venture firm, the foreign partner may take part in the management. For the foreign partners, it is a way to take care of their investment; and for the local partners, it is a way to absorb the experience of advanced management methods from abroad that are necessary for the successful operation of a modern production line.

Technology importation is also closely related to the technical and socioeconomic conditions in the importing country. In China, experience of technology importation has proven that the capability for assimilation and absorption plays an important role in the success of technology importation.

Transplanting a species from one place to another needs proper ecological conditions for growth. Likewise, the transfer of technology to a different country needs to adapt to local conditions. First of all, it needs to adapt to the local material supply. For instance, in the process of the technology assimilation and absorption of the imported 1.7 meter rolling mill, one of the key issues that needed to be addressed was the rolling mills mathematicals model. The original model introduced from Japan could only be applied to the rolling of steel made in Japan which used aluminum deoxidization, while steel made in China used silicon deoxidization containing two-thousandths copper. After the whole set of imported equipment was installed and thoroughly tested, the rolling of steel made in Japan went smoothly; but when it was used for rolling steel made in China, it happened that there was either a drawing breakage of steel or a trip of electric motor. The countermeasures adopted by the factory failed to solve the problem. Later on, the factory collaborated in research with university on the mathematical model for the rolling pressure. In the process of the research, faculty members and graduate students of the university, through a careful screening of multistructural forms of mathematical models, developed a model with minimum variance. By using such a model for the rolling of steel made in China, the best effects were achieved. This made it possible for a large imported rolling mill to operate normally and play a role in developing new products.

Technology importation also needs to outfit the imported technology with other links of the productive process. Sometimes, when carrying out projects of technology importation, people are used to focusing on the specific technology they are looking for. They take it as the key technology and believe that only by acquiring such technology can they make the operation work. However, in many cases, imported technology may fail not because of the incompetence of the core technology, but because the imported technology and other links of productive process do not fit together. For instance, in the mid-1980s, China imported about a dozen production lines for the manufacturing of zip-top cans. However, zip-top can manufacture required an aluminum sheet with high purity, as a single gas hole might spoil the whole can. As the aluminum sheet made in China was not up to standard, the material had to be imported. But, a hundred million U.S. dollars per year were required to procure the material from abroad. This price was not realistic for the zip-top can manufacturers in China, and as a consequence most of the imported production lines were left idle. Not until the aluminum process manufacturer in China imported updated aluminum filtering technology and improved the purity of the aluminum sheet to the standard for zip-top can production that most of the imported production lines of zip-top can were put into production. This process took a few years.

The success of technology importation depends not only on the fit of the materials with the imported production line but also on the match of imported technology with other links of productive process, and the adaptation of the products to the market. The products of imported technology have a certain life span and should continuously be adjusted, improved and renovated to adapt them to changing market demand. The importation of refrigerator technology is a good example. In the 1980s, China imported nine refrigerator production lines from the Ariston of Italy; the output of these production lines accounted for one-fourth to one-third of the total output of refrigerators in China. All nine firms, located in different cities throughout the country, bought the same production lines, making up an Ariston family of nine members. Eight of them added the foreign brand name Ariston after their city name or brand name, as in Meilin Ariston, Changling Ariston, and so on. Most of the nine production lines were imported in 1987 and 1988 during the boom in purchases of consumer electronic products in the country. This made the nine firms, as the owners of these imported refrigerator production lines, very profitable. After 1989, the market situation in refrigerators changed. By that time, about one hundred production lines of refrigerators, all over the country, had been put into production; the supply far exceeded the demand, and the business of the nine firms stagnated.

Five years later, after intensified competition in the refrigerator market, the nine firms were divided into three groups, according to their business records. Three of them formed the first group, which had joined the top ten of refrigerator manufacturers in China. Among the reasons for the gaps among the nine firms, one of the major reasons was innovation strategy. The most successful group, the Meilin Group, made just a minor change in the design: enlarging the capacity of the freezer section of the refrigerator. Though it was only a minor change, it was demanded by a large number of consumers and helped the firm to secure a large share of the market. It allowed the Meilin Group to move from 32nd place up to second place in the refrigerator industry. At the same time, another member of the nine firms also made a change, introducing a new design from Italy that doubled the space of the refrigerator. It turned out to be a failure, as consumers expected more freezer space instead of more refrigeration space. The case of the refrigerator shows that adapting products of imported technology correctly to changing market demand is a major factor in the success of importation.

With the capability of technology assimilation, absorption and innovation, the importation of technology served as a catalyst to promote not only the development of an individual technology or a specific product but also the growth of a whole new industry. The development of the color television industry is a good example. China imported the first color television receiver production line in 1980. That began the prosperous development of the color TV industry in China. From 1980 to 1995, China invested more than 15 billion yuan RMB in the color TV industry and initially formed a relatively complete color TV industrial system, including parts and devices, color kinescope and color TV receivers. To date, a production capacity of 20 million color kinescopes and over 36 million color TV receivers has developed. In 1995, the total output value of China's color TV receiver industry was 43.9 billion yuan RMB, accounting for about 18% of the total output value of the electronics industry in the same year. If the output value of parts and devices were included, the output value of the color TV industry would account for about 40% of the total output value of China's electronics industry. The color TV industry has been one of the mainstays in the field. In 1978, the output of color TV receivers in China was only 4,000 sets; in 1990, it increased to 10.3 million sets; and in 1995, it had broken 20 million sets and reached 20.13 million sets. China grew into a world power in color TV. The quality of color TVs made in China has improved constantly. New performances and new technologies available from the color TV industry abroad have been introduced into China very quickly. In addition, the ratio of performance to cost of color TVs made in China is superior to that of those made abroad. This is a result of adapting design to circumstances in China. Distinguished from the foreign design, which is created for a certain life span and will fail after that life span, China's color TV is designed for reliability and a life span as long as possible. The quality of color TV made in China is competitive with corresponding products made abroad. Therefore, there has been a substantial export of China's color TVs to many countries and areas in the world, including the United States, Japan, Europe, and Africa. In recent years, China has exported more than four million color TVs each year. Some of China's color TV manufacturers have set up overseas subsidiaries. Up to 1995, the total number of color TV receivers owned by Chinese families was 120 million sets, and the owning rate was up to 89% in urban areas and 17% in rural areas. A relatively complete broadcasting and TV program transmission and coverage network has been formed all over the country.

It can be seen in the development of the color TV industry that technology importation has played an important role in starting and helping

Table 8
Output of Various Durable Home Appliance Products in 1994

Product	Unit: in 10,000 Sets
Refrigerator	764
Washer	1,096
Air Conditioner	382
Freezer	234
Electric Fan	7,730
Vacuum Cleaner	171
Electric Cooker	1,645

Source: The Statistical Bulletin on the National Economic and Social Development of the PRC by the State Statistical Bureau.

speed up the development of an entire industry. On the other hand, it also shows the importance of technology assimilation, absorption and innovation. The development of the color TV is only one of many cases of technology importation. The entire durable home appliances industry, though having a rather late start, has developed at a relatively fast speed, reaching a total output value in 1994 of 74 billion yuan. In the past decade and a half, it increased by 30% per year, progressively. In 1994, the output of various products were as shown in Table 8. The output of the electric fan, washing machine, refrigerator and electric cooker manufacturers is the largest in the world, accounting for 10% of the total world output. Currently, the quality of major durable home appliances products is close to, or equal to, the international standard, and many have become brand name products. Some of the brand name products have been exported to a number of countries, including developed countries, and the hard currency earned from the export of these products has amounted to more than 700 million USD per year. Of course, success in the development of the durable home appliances industry is owed to the intensive investment made in the early stages of the development of the industry. However, technology importation and the successful assimilation, absorption and development of technology are also important factors for success.

In the 1990s, technology importation, through various joint ventures, has further expanded not only in scale but also in scope. After a successful joint venture with McDonnell-Douglas for the assembly of airplanes, joint ventures in the production of automobiles have developed rapidly. The importation of technology through joint ventures has expanded to more important industrial departments.

Mutual benefit is a principle of international cooperation, and the mutual supplement of each other's superiority is the basis for international cooperation. China needs to absorb foreign advanced technology and capital, and foreign businesses are looking for the broad market and lower production cost in China. In the 1990s, China has put forward a general guideline for the exchange of technology with market. For instance, one European company has a brand name product, a vibration electric motor, which is the core of vibration equipment. The product is of good quality and has had good market share in the world for a long time. However, in recent years, the company has suffered intensified competition from similar products from Japan and South Korea. Though the quality of the product of the European company is better, its price is much higher, because of high production costs. In China, there is a factory that produces the same product and has dominated market share in the country. However, because the core component is inferior, the life span of the vibration electric motor is much shorter. An excellent match would be possible between the two companies in China and Europe. The companies could set up a joint venture and transfer the core technology to upgrade the quality of the product, mainly the life span of the motor. After quality improvement, the product can sell at a better price in China; and after the production cost is lowered, the cheaper product will be much more competitive in world markets than it was before. Both parties of the joint venture will share in the results, but most important is the sharing of market and technology.

In the past two decades, as the investment environment has greatly improved, the capability of technology assimilation, absorption and innovation has constantly been enhanced. Especially, the mentality has greatly changed, and policies and regulations relating to opening have been further liberalized. Foreign investment and joint venture businesses have grown quickly. The growth of foreign investment in Beijing is a good example, as its evolution is typical in China. Foreign investment in Beijing started in 1979 with the establishment of the first joint venture hotel—the Jianguo Hotel. The establishment of the Beijing Jeep joint venture company was a breakthrough that marked the expansion of foreign investment into the industrial field. In 1992, the establishment of Yansha Friendship Shopping Center (a joint venture with Lufthansa of Germany) was another breakthrough, which marked the expansion of foreign investment in the commercial area. Since 1992, there has been a rapid growth of foreign investment in real estate: 1994 saw the first foreign investment in city infrastructure construction and 1995 was the year the first subsidiary of Tokyo Bank opened in Beijing. This last was a breakthrough in foreign investment in the banking area.

With the constant expansion of foreign investment into more economic fields, foreign invested joint ventures or foreign owned enterprises are distributed in 76 different fields of nine industrial departments, including agriculture, industry, construction, transportation, commerce, and real estate.

Regarding the scale of investment, there has been a constant increase in large- and medium-scale projects. Up to the end of 1995, there were 308 projects, with the total investment in each between U.S. $5 and 10 million, of which 75.6% were approved in the three years before 1995. There were 625 projects with the total investment in each one above U.S. $10 million, of which 80% were approved in the three years before 1995.

Regarding the technological level of foreign invested projects, there has been a constant growth of capital intensive projects containing higher technology. There were 91 joint venture enterprises identified as technologically advanced enterprises and 223 joint venture enterprises as export-oriented enterprises. All these enterprises have played an important role in the formation and development of the high and new technology industry in Beijing, thus helping improve the industrial formation and industrial level of the city. There are a number of backbone enterprises in different key industrial fields, such as large-scale integrated circuits, color kinescope, large-scale programmed exchanger, fiber optics cable, mobile telephone, jeep and light truck.

As for foreign investors, more and more world famous large corporations are investing in Beijing. By the end of 1995, 247 such business had invested in 307 projects, accounting for 2.4% of the total foreign invested projects, and their agreed investment amounted to 2.49 billion USD, accounting for 14.2% of total agreed investment. Most of the projects with an investment above U.S. $10 million were invested in by global corporations.

Since 1979, and in the 17 years up to 1996, foreign investment actually utilized by Beijing amounted to U.S. $8 billion, of which 72% was direct investment by foreign businesses, equivalent to 28.8% of the total local investment in fixed assets in Beijing. It has played an important role in the economic growth of the capital city. According to statistics, 70% of the large and medium-sized enterprises have used foreign investment at different levels. In recent years, more than 60% of the newly increased industrial output value has come from enterprises with foreign investment.

Since a number of relatively advanced applied technologies have accompanied the foreign investment, technology progress in Beijing has been accelerated and industrial upgrading has been promoted. A variety of industrial products in a number of industrial fields have been renewed. In comparison with the technological level of developed countries, prior to the development of joint venture enterprises, the technological levels of 80% of domestic enterprises remained at the level of the 1960s and 1970s; and now, when a large number of joint venture enterprises have been set up, the technological level of 63% of enterprises

Table 9
The Growth of Foreign Investment in China (in 100 Million Yuan)

Year	Foreign Capital Actually Used	Direct Foreign Investment
1992	188.0	111.6
1993	367.7	257.6
1996	552.7	423.5

Source: The Statistical Bulletin on the National Economic and Social Development of the PRC by the State Statistical Bureau.

has come up to that of the late 1980s, and 4% of the enterprises have come up to that of the 1990s. Henceforth, gaps between the city's industry and the international advanced level in technological development will be narrowed.

The development of joint venture enterprises has promoted the growth of export-oriented economy and helped to increase employment opportunities. According to estimates, since 1986, of the newly increased output value of exported industrial products, 45.07% has been derived from foreign investment. In 1995, the exports of the foreign invested enterprises grew to U.S. $665 million, accounting for 26.7% of the total export value of Beijing. The number of staff and workers employed by foreign invested enterprises was up to 330,000.

Obviously, foreign investment in Beijing has effectively helped promote economic growth. However, there are still a few problems.

Compared with the international advanced level, the technological level of the projects with foreign investment in Beijing is still low. There has been a repetition of projects introduced at lower technological levels. The situation of Pudong Special Economic Zone of Shanghai is in contrast to that of Beijing in this regard. In the past seven years, 70% of projects developed in Pudong are high technology projects. Pudong is taking the lead in carrying out a strategy of a high starting point and strict selection of projects with higher technological level. This is, to a great extent, owed to the favorable investment environment of Pudong, which is able to attract projects at higher levels. Nonetheless, in many cases, the situation in Beijing is more typical for the time being.

The experience of Beijing can be taken as a microcosm of the evolution of absorption of foreign technology and investment in China in the past two decades. Tables 9 and 10 provide an overview of the nationwide situation, showing that foreign investment in China has developed much faster in recent years. With the rapid growth of foreign invested enterprises, both the scale and scope of technology importation have had un-

Table 10
Growth of Import and Export of Enterprises with Foreign Investment (in U.S. $100 Million)

Year	Total Amount of Import and Export of Enterprises with Foreign Investment	Increase over Previous Year	Proportion in China's Total Amount of Import and Export
1995	1,098	25.3%	39.1%
1996	1,371	24.8%	47.3%

Source: The Statistical Bulletin on the National Economic and Social Development of the PRC by the State Statistical Bureau.

precedented development, while the import and export of enterprises with foreign investment have accounted for a substantial part of China's total import and export. However, quite a few problems have arisen from this development, especially when such development has accelerated. These problems give rise to a great deal of national controversy, related to the attitude, stance taken and policies formulated toward the absorption of foreign investment.

First, inasmuch as the scale of foreign investment is expanding rapidly and has grown to be a substantial part of China's economy, some experts are worrying about its consequences. They propose that certain measures should be taken to control the rapid development of foreign investment. The point has been refuted by other experts,[2] who make an overall approach to the issue and hold an optimistic view of the further absorption of foreign investment. According to their view, though China had absorbed a total foreign investment of U.S. $150 billion up to the end of 1996 and become the second largest foreign capital import country in the world, it was still far behind the United States, the first foreign capital import country, which had absorbed U.S. $2,300 billion by the end of 1991. In addition, China makes up a smaller share in the aggregate amount of foreign investment in developing countries and the countries that are transforming from a planned economy to a market economy.

People may ask if there is any limit to the absorption of foreign capital. It is difficult for anyone to set up a quantitative limit yet, but theoretically, there is some sort of limit. According to the viewpoint of some experts,[3] the limit of the economic development in a country is determined by whether the existing economic resources of the country are fully and effectively utilized. Economic resources cover the main factors of production. If a country's factors of production haven't been fully and effectively utilized, there should be a certain potential for economic development of the country. China is a large developing country and has a serious imbalance in the disposition of production factors. There is a large amount of idle labor and non-exploited natural resources in China. These idle production factors are the material basis for the utilization of foreign capital. Should idle natural resources and an idle labor force persist, foreign investment will continue to play a positive role in promoting the exploitation of the idle production factors and economic potential.

According to some experts, the question of limits may be answered from another aspect (i.e., the market accomodation principle). That is to say, if foreign investment can take a foothold in China or, in other words, be accomodated by the market in China, it should be taken to be a requirement of China's economic development. According to this principle, once foreign investment cannot be accomodated by China's market, the limit of absorption of foreign capital, will be reached.

In recent years, more global corporations have made inroads into the Chinese market in different ways. Aside from investing in China to set up joint venture enterprises, these corporations have merged with Chinese enterprises and made these enterprises their local manufacturers, set up a local manufacturing center directly or developed marketing and sales centers in China. In some industrial sectors, global corporations have tried to dominate market share by way of merging the local industries. According to statistics, in China's market, among the 10 world brand name cosmetics manufacturers, seven of them are in the market and make up 80% of market share. The foreign brand name machine tools make up a market share of 70%. Foreign telecommunications equipment has a market share of 80%. Some foreign consumer commodities are making inroads into China's market. For instance, foreign fruits are selling in China. Now, people in China are able to taste apples and oranges from the United States, the durian from Thailand, the crystal pear from Japan, and so on. In 1995, Shanghai sold about 20,000 tons of foreign fruit; and in 1996, sales were several times that of 1995. Even in Huhehut in Inner Mongolia, foreign fruit has spread from the large shopping centers to peddlers on the street. According to the figures provided by the General Customs Administration, the total volume of the fruit imported in 1996 was 645,000 tons at a total value of U.S. $196 million, increases of 172.0% and 134.9% over the previous year, respectively. In Shanghai, now it is difficult to get bananas from Guangdong, as they have been replaced by bananas from abroad. Last year, 13 million kilos of foreign bananas were sold in the market in Shanghai, equivalent to the volume of bananas from Guangdong sold in Shanghai over the previous few years.

As mentioned earlier, after developing for more than a decade and a half, China's color TV industry has a comparatively solid basis, with a large production capacity and technology development capability. China does have a large market, and it is exactly that large market that foreign color TV manufacturers are eager to exploit. The entrance of global corporations has led to intensified competition in the color TV market and brought about a serious challenge to the domestic color TV industry. For instance, in 1995, Japan had an annual output of color TVs of 49.65 million sets, among them only 723 million sets were made in Japan, while the rest of the products were made overseas, and most parts were made in Southeast Asian countries, including China. The large Japanese color TV manufacturers came to China one after another to set up manufacturing bases and established new joint venture enterprises with a large production capacity. For instance, Sony has developed a joint venture with Shanghai Broadcasting and Television Co. Ltd. and Shanghai Vacuum Electronics Device Co. Ltd. This joint venture enterprise is backed up by the advanced technology of Sony and will gradually introduce the large screen color TV technology, high-definition television technology, multimedia technology and high-resolution cathode ray tube technology with digital electronics technology as the core. The development center set up by the joint venture firm will, based on the advanced technology introduced from Sony, gradually form a technology development and new product development capability at world class level in the field of image technology. Meanwhile, Beijing National Kinescope Factory is planning to produce its National Color TV. While Japanese color TV giants are using the joint venture to penetrate China's market through production bases in Shanghai and Beijing, South Korea is developing a joint venture with Tianjin. Though China has a broad market, there is a limit to market accommodation. Now, China has a production capacity of more than 30 million sets, while the size of the color TV market in China in recent years has been 17 million sets per year. Apparently, China is going to suffer from a surplus of production capacity. The participation of foreign large manufacturers of color TVs can't but bring about more intensified competition and will make it difficult for a number of domestic color TV industries, which are at smaller production capacity and less well-known in their brand name, to survive. Nevertheless, difficulty coexists with opportunity. In China, every year there are 8–9 million newly married couples and the same number of new families; and at the same time, there are three and a half million to four million households moving into new houses. Many of them need to buy a new color TV. In addition, a majority of people bought TVs in the early 1980s, and now it is time for them to replace the old sets with a new color TV. Therefore, with the serious challenge, opportunities exist too. Insofar as the domestic color TV industry has had robust development and built up a relatively solid basis previously, it is still able to meet the intensified competition head on. The situation of China's soft drink industry is much more severe. In recent years, China's soft drink industry has developed rapidly. The total output volume in 1944 was 627.9 million tons, a growth 21.8 times that of 1980 and an annual increase of 25% progressively. Up to 1994, there were 2,827 soft drink enterprises, with 218,200 workers and staff and a total sales revenue of 11.787 billion yuan RMB. Intensified competition has taken place in the field of carbonated drinks, specifically, China's carbonated drink market has been occupied by the "two colas"—Coca-Cola and Pepsi. The "two colas" have experienced the typical process of foreign consumer products in China's market, that is, prohibited—restricted—permitted—deregulated. When Coca-Cola sent a set of their bottling production equipment to China as a gift at the end of 1979 to replace the importation of Coca-Cola for consumption in the international hotels in China, the Ministry of Light Industry rejected it out of worry about the impact on the domestic industry. However, the Ministry of Foreign Trade and Economic Relations accepted it and installed it at the warehouse of Beijing Grain and Edible Oil Import and Export Corporation for production to supply the international hotels. Later on, Coca-Cola was permitted to set up its bottling factories in a few cities. Now, the situation is quite different; the total annual output of Coca-Cola and Pepsi in China has risen to 1.11 million tons, accounting for 18.5% of the total soft drink output and 37% of the carbonic acid drink in China. After a few new bottling factories are constructed and put into production, the market share of the "two colas" will further expand. In China, there are eight brand name soft drink industries, and six of them have set up joint ventures with the "two colas." The "two colas" have the controlling share in most of these joint ventures, and all their main products and new products are using the brand names of the "two colas," except the old products, which keep their original brand names.

This situation, caused by the entrance of a large number of global corporations, certainly merits the attention of the domestic industry. The intensified competition has brought about debates both in theoretical circles and in the news media. Inevitably it will lead to calls for the protection of the domestic industry or the protection of domestic brand name products.

Some experts have made a few points about this issue. The economic development of a country certainly needs to rely on its own national industry. However, protecting national industry doesn't mean protecting a long-standing technological backwardness, long-standing low economic benefits and consistent monopoly over several decades. Development of national industry cannot depend on protection, but should rely on competition. National industry can only survive and get stronger with the availability of competitiveness. The competitiveness of an enterprise mainly manifests in its capability to earn hard currency through export, its level of productivity, per capita profit and tax payment and how large a market share it possesses. The industrial potential of a country doesn't lie in its protectionist policy or in high tariff trade barriers, but in its engineering technology, operational level and entrepreneurship.

China's automobile industry is a typical example of prolonged protection. The automobile industry, under protection at a high degree, after 40 years is still an "immature industry," while the television industry, under protection at a low degree, has an annual output ranking first in the world. It shows that for any national industrial sector, the higher the degree of protection, the less its opportunity in the market and the weaker its competitiveness.

Although some people take a positive stand for the further liberalization of the policy toward foreign trade and technology importation as well as absorption of foreign investment, some others are more concerned with the impact that foreign investment has on the improvement of industry formation in China.

Concerning the change of attitude toward foreign investment, people may find that in the early stages of opening up, attention was paid to the number of projects and the quantitative size of foreign investment. Now, attention has shifted to the distribution of foreign investment. In the 1980s, a large amount of foreign capital absorbed by China was invested in labor intensive industries. Many foreign invested projects were distributed among processing industries, and a few foreign invested projects were distributed among agriculture, energy, transportation, raw materials and the like; more foreign invested projects were distributed among primary processes and labor intensive projects, and fewer were distributed among high and new technology and technology intensive projects; many were distributed among small and medium-sized enterprises, and fewer were distributed among large enterprises. For instance, by the end of 1993, foreign investment in transportation and the telecommunications industry was 1.8%, in agriculture it was 1.54%, and in the energy and material industry it was only 0.1%.

The situation of foreign investment certainly relates to the improvement of China's industrial formation. In the 1980s, the development of the export-oriented economy did have its success economically; however, most of the export-oriented industries developed in the 1980s were labor intensive at a low technological level and with low added value. There was a sharp contrast between the low added value of exported products and the high added value of imported products and equipment. For instance, the product of the textile industry was one of the major items for export. The added value of textile products was as low as 1:1.4, while the added value of imported industrial sewing machines was 1:9.8 and that of imported metal processing lathe was 1:125. The large difference in added value between export and import placed China in an unfavorable position in international trade. The situation was aggravated by the high consumption of materials of the export-oriented industries, as these industries were both labor intensive and resource intensive. The incessant and rapid expansion of these industries inevitably led to tension between the supply of resources and agricultural products, as many industries had to use primary agricultural products as raw materials. The situation made sustained development impossible.

Such conflicts could be settled only by the upgrade of industrial formation and the level of the exported product mix. Therefore, a readjustment of domestic industrial formation was required. At the same time, the improvement of foreign investment composition is of significant importance.

The official stand toward the absorption of foreign investment will be significant. Has the absorption of foreign capital in China reached a limit, and is the adoption of certain steps required to control the further expansion of foreign investment? This question concerns many people, both at home and abroad. In his statement made at a seminar on foreign investment in China's economic technology development zones, Li Langqing, Vice Premier of the State Council, pointed out: "China's general policy is to use foreign investment in a more active and more effective way, so as to improve the composition and level." "More active" means to continue to absorb more foreign investment in a better way. According to Li, this is an irreversible tendency. "More effective" means to effectively direct foreign investment according to State industrial policy, develop more projects at a higher technological level and integrate foreign investment with the development planning of the national economy in an organic way, to obtain the effect of mutual benefit.

After the Fifteenth Congress of the CPC, both China's foreign trade and China's use of foreign funds have come to a new stage.

The period of the five years since 1992 saw the fastest growth of foreign trade. During 1992–1996, the total amount of China's foreign trade came to U.S. $1,168 billion, and the growth rate of China's foreign trade per year was 16.4% on average, which was not only higher than that of China's national economy in the same period of time, but also 8% higher than that of world trade. It is estimated that by the year 2000, the total amount of annual import and export will come to $U.S. 400 billion, which means that from late 1997 to the year 2000, China will provide the world with a market of U.S. $700 billion and a total trade opportunity of U.S. $1,400 billion.

According to the latest news, China is planning to make a total investment of about U.S. $750 billion in capital construction in the next three years. A few figures have been made public that show quite a few large-scale capital construction projects will be under way in the coming years. It includes 80 million kilowatts of newly installed power capacity in power construction, 16,000 kilometers of newly constructed railway line, 2,800 kilometers of newly constructed superhighway, 150,000 kilometers of newly installed fiber optics cable and the increase of telephone exchanges to a capacity of 80 million new subscribers. It means that substantial foreign funds will be used for large-scale capital construction projects.

Apparently, the entrance of global corporations into China's market has brought about a strong impact on domestic industries in quite a few fields. However, it doesn't mean that China is going to adopt a protectionist policy, as some people expect. As the opening policy is a State strategy that covers the long-standing and overall interests of the country, it is not likely to be impacted by temporary interests of certain specific industries. According to the official statement, China will further cut down custom tariffs, causing the average cost of customs to decrease by 17–23%, a big reduction. China promises to abolish all non-tariff barriers that do not conform with the agreement of the World Trade Organization in 15 years. In the past five years, China has already expanded opening up from the coastline area to the broad hinterland and formed a layout of opening up on all sides and on multiple levels, including coastline areas, the areas along the Yangtze River, border areas and various inland provincial capital cities. With the further adoption of measures on a joint track with international conventions, there will be no room for protectionism. China has an old idiom: "Don't give up eating for fear of choking."

As mentioned earlier, the opening policy, including the absorption of foreign investment, foreign advanced technology and management experience, is an important part of the development strategy of China and represents the long-standing and overall interests of the country. Obviously, it would not be wise to give up such interests just for fear of the intensified market competition derived from the entrance of global corporations. However, the opening policy doesn't mean that no support will be given to a limited number of industries, for a limited period. Support will be given to industries with prospects and markets that may grow stronger with competitiveness in intensified market competition; in other words, support will not be given to backward industries without prospects. "Select the superior and eliminate the inferior" is the law of market competition. To protect backward industry without prospects will distort the principle of a market economy and make domestic industries depend on protection, instead of their own competitiveness. The best way to protect the national industry is to encourage domestic enterprises to try in every way to remold themselves and grow into modernized enterprises with competitiveness.

Upgrading industrial formation is necessary for China's industrial modernization as well as the development of the national economy. This is exactly the purpose of absorbing foreign investment, advanced technology and management experience. During a certain period of time, especially during the early stages of development, the development of labor intensive export-oriented industry was necessary, and such industry will continue to play a role in China's national economy for a period of time. However, labor intensive industry in China will inevitably and gradually be replaced by technology and capital intensive industry. Right now, there is a process of shifting labor intensive industry from the coastline areas to inland areas. It is a result of the increase in labor cost, the lower added value of labor intensive products and the limitation of the export market. However, during the early stages of development, much of the foreign investment was in labor intensive industry. Now, in the process of upgrading industrial formation, the original composition of foreign investment turns out to be disproportionate. In order to turn around this misplaced investment, during the ninth five year plan, the government is trying to strengthen the policy guiding foreign investment in accordance with the guidelines of the government industrial policy.

During the period of the ninth five year plan, a series of reform and policy adjustments will be gradually carried out in respect to foreign trade and foreign investment, including the implementation of national treatment[4] of enterprises with foreign investment, the cancellation of preferential treatment for enterprises with foreign investment, cutting the import customs tariffs,[5] and the expansion of the experiment of the BOT (build, operate, transfer) investment form.

The policy adjustment is mainly for the purpose of reaching a joint track with the international economy. The adjustment will help place market competition on a fair basis. Though the cancellation of preferential treatment will affect the interests of foreign investment, the national treatment will actually gradually cause foreign investors to enjoy the same treatment as domestic enterprises in terms of banking debts, financing, investment and a unified business income tax rate. The impact on enterprises with foreign investment of the cancellation of preferential treatment in customs on imported production facilities will be partly offset by the reduction in the import tax rate.

Absorbing foreign investment, advanced technology and management experience has promoted China's economic development and reform; in turn, furthering current economic reform will play an important role in promoting China's further absorption of foreign investment, advanced technology and management experience. There is an interaction between the two.

China's industrial reform is significant in furthering industrial strength and competitiveness. There are two major aspects of the reform: (1) promoting the formation of a modernized enterprise system and (2) promoting a scale economy through merging and conglomeration of industrial enterprises.

According to Jiang Zeming's statement in his report at the Fifteenth National Congress of the CPC, the establishment of a modernized enterprise system is the direction of reform for State-owned enterprises. It is necessary, in accordance with the requirements of "the distinction of property right, the clearness and definiteness of power and responsibility, the separation of government administration and enterprise and the scientific way of management" to carry out the reform of State-owned large and medium-sized enterprises according to the normalized corporate system, to turn enterprises into legal entities and the main bodies of market competition.

The promotion of the share holding system is an important way to solve the problem of property rights. In China, the lack of vigor of State-owned enterprises is fundamentally derived from the absence of clear-cut property rights for enterprises. Who is the owner of the enterprise? Who is responsible for the profit and loss of the enterprise? This has always remained a serious problem. Previously, the answer was that since all State-owned enterprises were owned by the State, the owner was the State. However, it is difficult to materialize the State ownership in each specific enterprise. Of course, it was the director or the manager of the enterprise who took responsibility. However, the director or the manager of enterprise was appointed by the government department in charge, and all major issues were reported to and approved by the government department in charge. Thus, the director or the manager could only take limited responsibility. If State-owned enterprises were able to continue their normal operations under a planned economy, it would be very difficult for these enterprises to survive under the new conditions of a market economy. As State-owned industrial enterprises have been used to a product economy or a quantitative economy, they cannot develop themselves in a market economy environment. The government tried various ways to improve the situation of State-owned enterprises, such as to separate the ownership and operation by carrying out a contracted responsibility system, but all these measures did not turn around the situation. With the growth of marketization and intensifying market competition, the situation led to a heavy deficit in nearly half of the State-owned enterprises and a heavy burden for the government. According to statistics, State capital that could be used for the production and operation of State-owned enterprises was only 3,000 billion yuan, but it was distributed among 291,000 industrial and commercial enterprises, ranging from long-range missiles to small catering businesses. As people described it, it was like using ten fingers to squish a hundred fleas.

An ancient poem goes: "By the dead end of the mountain and river, it seems there is no way out; the dark and thick willows and bright blossoms point to a new village." The new village ahead is the share holding system. China has worked out a new system of corporate law, which already provides a legal basis for the formation of modernized enterprises. The remolding of State-owned enterprises into share holding companies will help separate the enterprises from government administration. The share holding system will diversify ownership, as the shareholders may be the State, individuals, management teams, workers and staff, or even different enterprises holding each other's shares. The diversified groups of people as shareholders will take care of their interests in the enterprises and elect a board of directors to direct the enterprise. The State, as an important shareholder, will have State asset management companies to protect its interests, and the latter may appoint their members to sit on the board. Therefore, property rights will be distinctly clarified.

As a result of the original central planning economic system, China's enterprises have been limited to a smaller scale and industrial strength has been dispersed. For operation under a planned economy, this was acceptable, as all enterprises under a specific industrial department were administered by it, and the departmental administration could coordinate its various enterprises. Industrial linkages are realized by department administration under a planning economic system in China. That's the reason why on the list of the world's 500 largest corporations, not one belongs to China, although China is able to match its total production capacity with countries that possess the world's largest corporations in industrial sectors such as iron and steel, coal, petroleum, chemicals, building materials and color TVs. In 1994, China ranked first in the world in the production of coal, cement, cotton cloth and television sets and second in steel products, chemical fiber and chemical fertilizer. Many of the world's largest corporations are based in these industries.

Now, with the growth of a market economy, limitation to a smaller scale restricts enterprises from taking part in both domestic and international competition. Currently, a number of enterprises are carrying out mergers and forming a group of very large companies. The process of conglomeration is under way. For instance, a large petrochemical group has been formed. In Jiangsu Province, there were five petrochemical enterprises, namely, Jinling Petro-chemical Company, Yangtze Petrochemical Company, Yizheng Chemical Group, Nanjing Chemical Group and Jiangsu Petroleum Group Ltd. Now, the five companies are merged to form Donglian Petro-chemical Group, with total assets of 54.2 billion yuan and 145,700 workers and staff. Another example is the merger of airlines. A few years ago, after the deregulation of civil aviation, a number of local aviation companies were established to replace CAAC's[6] role in running the business of commercial flights. Deregulation was intended to promote competition among companies, to make it possible for civil aviation in China to improve its management and service. However, the dispersion of aviation companies decreased each company's business and reduced competitiveness. Now, International Air Flight, based in Beijing, is going to merge with Shanghai Eastern Air Lines, Hainan Airlines and Shangdong Airlines to form an aviation industrial group.

The reform and upscaling of enterprises will help to further the competitiveness of large and medium-sized enterprises in China and improve their status in intensified market competition. It is a better way to help domestic industry than protectionism. In turn, with a number of domestic industries as stronger competitors, foreign competitors will be stimulated to bring in more advanced technology and higher level products to upgrade China's industrial formation. International economic cooperation always accompanies international competition.

The reform of China's State-owned enterprises will also help technology transfer. Technology transfer needs not only a legal environment, but also a healthy business environment. Instead of a number of enterprises under government administration, a large number of independent modernized corporations will be the basic units of the economy and society. Each corporation will have its own independent interest to protect. Reform will provide good protection for the intellectual property rights of the foreign transferor. In the past, under the unified department administrative leadership, there were no distinct property rights for individual enterprises, and an enterprise was not able to distinguish its independent interest from those of other enterprises under the same industrial department. Therefore, even though there was a series of distinct legal provisions, there was no safe protection against the extension of know-how or technology knack beyond the limit agreed to by the transferee. Now, the counterpart of the foreign technology transferor will be a shareholding corporation with an independent business interest, which would counter any leakage of secret technology. Therefore, the intellectual property rights will be under double protection—the legal protection and the protection of the mutual interests of the transferor and transferee.

With the reform of State-owned enterprises and the formation of a modernized enterprise system, there will be a breakthrough in the development of marketization. The separation of enterprises from government administration will help withdraw government intervention both vertically and horizontally. In the past, State-owned enterprises were either under the direct control of the concerned central ministries (vertical control) or under the control of various local administrations (horizontal control). Each department or locality had its specific departmental or local interest, which led to various special rules and measures for the protection of departmental or local interests. This situation has played a part in partitioning the market. Only by withdrawing both departmental and local administrative intervention can a unified national market come into being. In the meantime, the commodity market and labor market will be improved. The reformation of financial markets will develop much faster than before, as the financing of a shareholding corporation will directly rely on financial markets, instead of depending on bank loans backed up by government policy. The stock market will be enlarged and play a more important role. The production factors market (mainly the property right transfer market), which has already emerged in a few large cities, will have rapid growth, as the share holding system requires an invigorated property right market for asset transfer. The furtherance of marketization can't but encourage the widening of foreign investment.

Last but not least, the reform of State-owned enterprises and the formation of modernized enterprises will provide more opportunities for foreign investors to pursue their best counterparts. The large and medium-sized enterprises with better production facilities and a technical force on a large scale will be open to various kinds of cooperation with foreign businesses. Previously, although these enterprises, even when bound by certain administrative oversight, had entered into cooperation with foreign businesses, there were a number of limitations. In the future, these enterprises, as independent modernized enterprises, will be prepared for business arrangements that follow international business conventions. With the availability of better production conditions, business cooperation between these enterprises and foreign corporations will be carried out on a larger scale and at a higher level.

NOTES

1. "Fengqing" is the name of a ship built in China. When it was finished at the later stage of the Cultural Revolution, the "Gang of Four" used it as an excuse to repudiate "comprador philosophy" in an attempt to attack Deng Xiaoping. Later, this was called "Fengqing Ship Event."

2. See Hu Angang, "Is There an Over-absorption of Foreign Capital in China?" *International Economic Review*, Nos. 3–4, 1997. The article has a straightforward exposition on trade liberalization.

3. See Zhao Xiuchen, "Talking about Some Problems on Using Foreign Capital," *Future and Development*, No. 1, 1997. The article has an in-depth analysis of the necessity of using foreign capital.

4. National treatment stands for a fair treatment to enterprises with foreign investment. The preferential treatment previously offered to the enterprises with foreign investment will be cancelled, and, in turn, these enterprises may enjoy the same bank debts and other treatment on a equal basis with domestic enterprises.

5. China has recently carried out several rounds of tariff cuts with a view to gradually achieving the requirement of the WTO. For instance, in November,

1993, the import customs duty rates of 2,898 taxable items were adjusted. In 1994, the import customs duty rates of 234 taxable items were reduced. In December, 1995, the State Council announced that beginning April 1, 1996, the average level of import customs duty would be reduced to 23%. China's customs duty was further reduced after 1996.

6. CAAC is the abbreviation of Civil Aviation Administration of China.

9

Readjustment of Science and Technology Policy in the 1990s

With the development of economic as well as scientific and technological reform, readjustment of the science and technology (S&T) policy was put on the government's agenda in the 1990s. Discussion regarding the new phase of scientific and technological reform had been going on for a few years. There was a consensus that the implementation of the S&T policy formulated in the 1980s had achieved its main goal. It fostered the concept that science and technology should be oriented toward economic construction among the broad masses of scientific and technical personnel in the country and helped initially to turn around the divorce between science and technology and production. The channels for integrating science and technology with economic construction had been created. However, the establishment of a socialist market economy in the 1990s raised new demands for scientific and technological progress. Insofar as the model of economic development in China is to transform from quantitative growth to growth in quality and economic benefits, science and technology is destined to play an ever more important role in economic development.

THE BACKGROUND FOR THE READJUSTMENT OF S&T POLICY

In 1995, a formal decision was made by the Central Committee of the CPC and the State Council concerning the acceleration of progress in science and technology. In 1996, a new decision was issued by the State Council on deepening reform of the science and technology system during the ninth five year plan, based on principles set forth in the previous decision and making it more specific. Both decisions virtually set up a new course to further the development of science and technology in China. Of course the new S&T policy is based on that of the 1980s, but from limited experience, what the S&T policy provided in the 1980s was a general direction, while many of specific forms and methods arose from the subsequent practice. For instance, non-governmental science and technology enterprises were first created by scientists and engineers themselves and only later on received the support of senior authority. Practice not only brought about positive experience but also provided negative experience from which people could draw valuable lessons. For instance, a low profile was taken toward the development of basic research in the 1980s, which weakened basic research in a certain number of important fields for a rather long time. Likewise, the policy issued in the 1980s took a lower profile in the development of high technology.

In the 1990s, the situation has changed greatly. There has been rapid economic growth as well as growth in people's demands for an improved standard of living.

China participated in international economic exchange to a greater extent than ever before and was confronted with strong international competition; the entry of a large number of global corporations couldn't help but intensify such competition.

Although low technology and labor intensive products had helped China develop a good export market, now the market for such products is saturated. In addition, exporting such products in large volume is equivalent to exporting resources and energy. The high consumption of resources and the increase in labor cost made it impossible to sustain the export of such products for a long period of time. For instance, in the past, China exported to Japan every year about 20 billion pairs of disposable chopsticks, which accounted for the entire consumption of chopsticks in Japan. Disposable chopsticks as a commodity have a very low added value and a high timber consumption. China is seriously short of forestry resources, with a vegetation rate as low as 14%. The vegetation rate of Japan is more than three times that of China. Actually, the importation of disposable chopsticks had helped Japan preserve its forestry resources even though its vegetation rate is much higher than that of China. In addition, Japan had recycled the used chopsticks into raw materials for making recycled paper. What Japan received from exporting this recycled paper had nearly covered its payment for importing disposable chopsticks. Therefore, for China's industrial upgrade, the export industry of chopsticks should be eliminated.

The upcoming participation in the World Trade Organization (WTO)

Table 11
Survey on 1,180 Sets of Equipment Owned by Large and Medium-Sized Enterprises

Up to International Standard	26.1%
Up to Domestic Advanced Level	27.7%
Up to Domestic Average Level	33.4%
Outdated Domestically, to Be Eliminated	12.8%

Source: According to China's Third Nationwide General Industrial Survey by the State Statistical Bureau.

means, sooner or later, weakening the edge of protection. This change is being brought about and will be furthered by the reduction in tariff rates and the withdrawal of non-tariff barriers.

All the aspects just mentioned boil down to one point: the readjustment of industrial formation and the enhancement of technology for industrial products have become pressing issues in improving China's international competitiveness. Under such circumstances, a higher profile has to be taken in the new S&T policy in terms of the development of high technology and high-tech industry.

Even more important is recognizing the interaction between scientific and technological development and economic mechanism. Stagnation of the large and medium-sized enterprises has been a negative factor for scientific and technological development in the industrial sector. Without strong internal demands from large and medium-sized industries, it has been difficult to put some of the major policies of science and technology into practice. Now, people have a sober recognition of the dependence of scientific and technological development on the economic mechanism. Both of the recent decisions have explicitly named the transformation of the economic growth model and the realization of the modern enterprise system as the premise of further scientific and technological reform.

On the other hand, realizing the transformation of the economic growth model is dependent on scientific and technological progress. The economic growth at lower quality is related to the lower scientific and technological level. Even though China has done a lot since the implementation of reform and opening policy in industrial technology renovation and has imported a large number of advanced technologies and equipment from abroad, the current technologies and equipment are still generally obsolete. Table 11 is drawn from a survey by the State Statistical Bureau in 1996.

Aside from the obsolescence of technological equipment, higher consumption of materials and energy as well as lower quality in products also come from the lower levels of technology.

The ability to innovate is one of the major indexes in judging the technological level and competitiveness of industry. In China, many new technologies are still being introduced from abroad. The lack of innovation in China's initiatives reflects the lower technological levels and competitiveness of her industry. It is no wonder that, under these circumstances, the economic benefits from industry have been difficult to improve. Therefore, the readjustment of S&T policy in the 1990s is centered around the improvement of economic benefits by enhancing the scientific and technological level.

THE GOAL OF THE READJUSTED S&T POLICY

The decision of the Central Committee of the CPC and the State Council pressed for a strategic transformation of economic growth from the quantitative growth model to a model of the growth of quality and benefits. This strategic transformation must rely on progress in science and technology, on liberalizing and developing the first forces of production, on strengthening the transformation of technology results into a realistic productive force, on transforming economic construction onto a track of relying on scientific and technological progress and on improving the quality of the labor force in a practical way. In view of this, the Central Committee of the CPC and the State Council decided to implement a strategy of "to rejuvenate the nation by science and education" with perseverance.

The Decision points out: In order to realize further liberalization and development of production forces in science and technology, the reform of the scientific and technological system must be deepened, to give access to initiatives and the creativity of the broad masses of scientific and technical personnel and to mobilize the entire social force to promote the progress of science and technology all around. The goal for the year 2000 is initially to set up a scientific and technological system that is to adapt to the socialist market economic system as well as the laws of science and technology, to make major progress in industrial and agricultural scientific research and technology development and in basic research and high technology research, to notably enhance the rate of contribution of scientific and technological progress to economic development, and to allow economic and social development to transform to rely on scientific and technological progress and on improving the quality of the labor force. The goal for the year 2010 is to further consolidate and improve the newly established scientific and technological system and to realize the organic integration of science and technology with the economy, to bring about prosperity through science and technology and to bring the scientific and technical force to a high level. The scientific and technological, as well as cultural, quality of the whole nation shall improve notably. The goals also include achieving an advanced level internationally in scientific and technological strength in major disciplines and certain high technology fields, improving the capability of innovation of China's own initiative by a big margin and mastering key technologies and systems-designing technology in major industrial fields, achieving or accessing the level of developed countries in manufacturing technology in the main fields by the beginning of next century, and achieving international advanced levels in manufacturing technology of certain rising industries. These achievements are meant to lay a solid foundation for a socialist modernized power.

HIGHLIGHTS OF THE READJUSTED S&T POLICY

High-tech industry is in the forefront of international economic, scientific and technological competition. According to the decision of the Central Committee of the CPC and the State Council, the development and industrialization of high technology is a fundamental way to upgrade industrial formation and improve productivity and economic benefits by a huge margin. Therefore, priority should be given to the development of high-tech industry in State industrial policy and development planning. Efforts should be made to improve performance, quality and market competitiveness of domestic high-tech products and to increase the scale-up benefits of high-tech industry and its proportion in the national economy, to make it possible for high-tech industries to gradually grow into mainstay industries of the national economy. It is necessary to have the development of high technology closely integrated with the demands of the national economy and defense construction. It is important to be reminded of the world tendency toward high-tech development, to insist on "focusing on limited targets and giving prominence to focal points" and to improve the capability of innovation of China's own initiative, enhance economic competitiveness, master intellectual property rights and realize industrialization.

With respect to developing basic research, State targets should be given a high priority in formulating a research plan. The central task for basic research is to provide a driving force for national economic and social development. Basic research should be focused on settling basic theory and technological problems as well as developing new technologies and methods for future economic and social development. It is necessary to emphasize the development of new leading disciplines, borderline subjects, interdisciplinary subjects and applied basic disciplines, to support the integration of natural science with social science and to pay attention to soft science research and application. The Decision also points out that for basic research, it is necessary to adhere to the principle "to grasp on something instead of everything," to aim at State targets and the world scientific frontier and to explore boldly and to be brave in innovation, in order to strive for raising science to new heights.

With regard to optimizing the formation of scientific and technological organizations and the distribution of scientific and technical personnel, the Decision set up a principle: "to consolidate one side and to invigorate the majority." "Consolidating one side" means taking the government's input as the rule and consolidating a small number of key research institutions and institutes under universities that engage in basic research, or applied research relating to overall and long-range interests of the State, or high technology research and social public welfare research, and tackling major scientific and technological issues. "Invigorating the majority" means liberalizing and invigorating many technology development and service institutions that have close links with economic construction and letting these institutions integrate with the economy in multiple forms and diversified channels. This dual principle was first put forward by Jiang Zeming.

THE STRATEGIC MEASURES OF THE READJUSTED S&T POLICY

The decision issued by the State Council on the Deepening of Scientific and Technological System Reform during the Ninth Five Year Plan was based on principles set forth in the previous decision but made it more specific. A few specific strategic measures for implementing the new S&T policy were put forward by the decision of the State Council.

The first strategic measure was to propel research institutions to become involved in the main battlefield of economic construction. Economic construction or industrial and agricultural production should be the main battlefield of scientific research. The majority of the scientific and technological force in the country should be directly involved in this main battlefield. However, in the past, such efforts were misplaced. For instance, the proportion of R&D expenditure of industrial enterprises in China was 30% and the proportion of average industrial R&D expenditure in total sales was only 2–3%. Scientists working with industrial enterprises accounted for only 25%. In China the government, rather than the industrial enterprises, has played a central role in scientific and technological research. Government-directed research not only led to the divorce of research from production but also caused low efficiency in research. In China, because of limited financial resources, R&D input is at a very low level—0.5% of GNP. It is at an even lower level than in some other developing countries, let alone comparisons with developed countries. With this limited input, the efficiency of scientific and technological research is at a lower level. The government-directed S&T research mechanism is one of the major reasons for low efficiency in industrial research. Departmental partitions and barriers derived from the government-directed research mechanism also cause low efficiency. Transforming government-directed research into industry-directed research is key to further integrating research with production and improving the efficiency of scientific and technological research.

The decision of the State Council, based on past practices, provides a blueprint for establishing a new systematic scientific research system that will involve the research force in the main battlefield in the country. The following three points were made.

1. An overwhelming majority of research institutions should take economic construction as the main battlefield and readjust their composition of specialties to contribute to remolding conventional industries, improving the quality of products, developing national industry, enhancing international competitiveness and strengthening comprehensive national power.

Research institutions with adequate conditions may directly join enterprises and become the technology development institutions of enterprises. The research institutions that serve various industrial sectors may adopt a membership system or a stockholding system and become the technology development institutions of each industrial sector.

Most of the technology development and service institutions should prepare to carry out business management, to enter markets through various forms and to develop themselves in market competition.

Research institutions with superiority in R&D and self-development capability or industrial development strength may start up their own businesses or directly turn to businesses. These businesses may be firms that integrate research, development, engineering design, production and marketing; they may also, by merging or contracting with other enterprises and research institutions, grow into enterprise groups and, through transferring technology with compensation, contracting technology development projects and providing society with special technology services, operate and develop on their own initiative.

Research institutions with the capability of both system development and technology development may keep close linkages with enterprises and transform into engineering centers or start up their own engineering centers.

Research institutions with a stronger comprehensive service capability may transform into a production force promotion center or technology innovation extension center and carry out business operations that offer technology services for society.

2. In order to accelerate the process of agricultural modernization, the agricultural R&D system and technology extension service system must be strengthened. To transform scientific and technological results, it is necessary to optimize the composition of discipline specialties and qual-

ified personnel for agricultural research institutions and to pay attention to multidisciplinary and transdepartmental approaches to key issues. It is also necessary to set up a number of State, trade and regional research and development centers to gradually form a dual-level key agricultural scientific research system, on both the central level and the local level. The State agricultural research institutions and key agricultural universities should concentrate on basic research and nationwide major applied research and development projects that solve major scientific and technological problems with overall, basic and directional importance. It is inappropriate that this be undertaken by provincial and district agricultural research institutions. Provincial and district agricultural research institutions should gradually develop into regional agricultural research and development centers according to natural and economic divisions and focus on carrying out applied technology research, follow-up development work and the transformation of scientific and technological results. Agricultural research institutions at various levels, when "consolidating one side," should make an effort to organize scientific and technical personnel to take part in developing science and technology demonstration zones, carrying out agricultural comprehensive development and programs that help the poor through science and technology and various scientific and technological services. It is essential to promote the integration of agriculture with science and education, to improve the scientific and cultural knowledge level of the peasants and their ability to apply agricultural science and technology and to train and stabilize agricultural technology extension workers in multiple skills in order to improve the overall agricultural productivity. It is necessary to improve and strengthen the agricultural technology extension system in a practical way, to consolidate and improve agricultural technology extension organizations at the grassroots level and to strengthen their self-development capability. The agricultural technology extension organizations should work in cooperation with various rural service organizations to form scientific and technological service networks to integrate technology, agriculture and trade and a connected sequence of production, supply and marketing. The agricultural technology extension organizations may engage in business activities in connection with their professional work by running business on their own or through combined business operations with seed companies or farming materials companies, and the revenue will be used to support the agricultural technology extension work and to pay stipends for the technical staff.

3. Enterprises should form the main body of technological development. Large and medium-sized enterprises should use the establishment of technology innovation mechanism of enterprises and the improvement of technology innovation capability as important content in estab-

lishing a modern enterprise system. Large and medium-sized enterprises and enterprise groups should be market oriented and gradually set up technology development institutions in combination with universities and research institutes in various forms. It is also necessary to encourage the mass technology innovation activities to increase technology input, to accelerate transforming and extending scientific and technological results, to develop new products and to support medium-sized and small enterprises and township enterprises in setting up long-term stable cooperative relations with research institutions and universities. It is also necessary to encourage the development of non-governmental science and technology enterprises.

The second strategic measure was to develop high technology and promote the industrialization of high and new technology.

High technology R&D is the vanguard of the development of a modern economy and the source of the development of the high-tech industry. It is necessary to select key issues important to the development of the national economy. The selected issues should be in the fields in which China has a certain foundation and superiority. Research results on these issues should strengthen the comprehensive national power of the country. It is necessary, through competitive bidding, to organize and promote universities, research institutes and enterprises to tackle these issues with joint efforts. It is essential to consolidate a high-level scientific research force to develop high-tech results and realize industrialization, especially that which is used in the readjustment of the industrial formation and technology renovation of large and medium-sized enterprises.

It is necessary to further promote new high-tech development zones. Reforms should be expanded to create a better environment for attracting research institutions, universities and large and medium-sized enterprises. This will help to develop high-tech industries and to accelerate the pace of reform according to the requirements of a modern enterprise system. Universities should be encouraged to start and run high-tech industries in the direction of the reform of higher education institutions.

The third strategic measure was to improve the formation and layout of research institutions that perform basic research. To carry out basic research, it is necessary that research institutions under the central departments and key universities serve as the main force to strengthen their joint efforts. It is necessary to consolidate key research institutes for basic research and place them under good direction to strengthen the basic research base and infrastructure, in order to choose capable scientific and technical talent to engage in basic research and bring along young talent. Such institutions should be optimized in organizational structure, and personnel should be diverted to improve efficiency.

It is necessary that basic research faculty participate in a two-way flow and accept concurrent appointments mutually. Effective measures should be adopted to promote joint development and the sharing of large-scale instruments and equipment, materials and information.

10

China's Industrial Policy and Technology Acquisition

Technology policy is closely related to industrial policy, but each has its special emphasis. Industrial policy focuses on the priority of the order of development of various industrial departments in the national economy, the proportion of development of different industrial segments and the priority and proportion of development within each industrial segment. All these should be based on China's national conditions and State strategic economic development goals. Technology policy mainly emphasizes the selection of technology and the focus of technology development under the direction of State industrial policy.

In 1986, the State Council issued the main focal points of technology policy for 12 sectors, including energy, transportation, telecommunications, agriculture, consumer products, machinery, materials, building materials, city construction, urban and rural home-building and environmental protection. The technology policies of the 12 sectors were openly published in 1988 for implementation.[1]

With the furtherance of economic reform and the development of the national economy, both industrial policy and technology policy were revised and given new emphasis. In 1989, the State Council issued the "Decision of Industrial Policy." In 1994, as China's reform and opening up, as well as modernization, reached a new stage, the State Council issued the "Outline of State Industrial Policy during the Nineties," which put forward the chief principles of readjustment and optimization of industrial formation, focal points of industrial development, and corresponding policies in order to ensure implementation.

During the 1990s, China was in a stage of accelerating the establishment of a socialist market economy; therefore, the market played a more important role in the disposition of resources. At the same time, the State was strengthening its macro control of the long range rational disposition of resources. As China was establishing her market economic system while the country was experiencing great change in industrial formation, the government was required to formulate a rational and predictive policy of industrial formation in order to consciously direct economic development and avoid errors in production and investment.

Industrial policy played a role of overall guidance for industrial development, and technology policy should be observed in light of the experience in industrial policy. It is, therefore, necessary to give a relatively detailed account of China's industrial policy before describing technology policy for various industrial segments.

With regard to the general objectives of the industrial policy formulated by the government, the "Outline of State Industrial Policy during the Nineties" points out, "The formulation of industrial policy serves as an important means to strengthen and improve the macro-control, to effectively adjust and optimize the industrial formation, and to improve the industrial quality, so as to promote the sustained, fast and healthy development of a national economy."

With regard to the major issues to be dealt with by the State industrial policy in the 1990s, the Outline makes a few points, as follows:

- to constantly intensify the basic status of agriculture and develop agriculture in an all-round way;
- to make big efforts in strengthening basic industry, to strive to mitigate the serious lag in infrastructure and basic industry;
- to accelerate the development of mainstay industries, to bring along the all-round revitalization of national economy;
- to reasonably readjust the composition of foreign trade, to strengthen the competitiveness of China's industry;
- to speed up the pace of development of high and new technology industry and support the development of rising industries as well as the development of new products; and
- to continue to make progress in developing tertiary industry.

In the meantime, according to the State Industrial Policy, it is necessary to optimize the industrial organizational structure, to upgrade the industrial technological level and to make industrial formation more reasonable.

DEVELOPING AGRICULTURE TO INCREASE PEASANTS' INCOME

It is significant that the Outline makes the development of agriculture the first priority of economic development. It points out that it is essential to set up a concept of "macro agriculture," which means that it is necessary to develop agriculture, forestry, animal husbandry, agricultural side products and fishery in an all-round way. The outline demands that strong efforts be made to develop an agriculture with high yields, fine quality, high efficiency, the capacity to earn hard currency, and match the quantity, varieties and quality of agricultural products to the demands of a well-off living standard for the people all over the country as well as the demands of the international market. It is necessary to readjust rural industrial formation reasonably, to actively develop the second and tertiary industries in the rural area and to gradually transfer surplus labor force in the rural area, in order to uplift the comprehensive productive capability and economic benefits of the rural area to a new height.

The significance of the principles should be understood in contrast to past experience and lessons. During the 1950s, when the development of heavy industry was made the first priority in economic development, the importance of the development of agriculture was overlooked; the consequences were serious in a country with such a large population. Later on, the order of priority was reversed, to become agriculture, light industry, and heavy industry. However, in the implementation of this policy, people weren't able to adhere to the principle with perseverance. Time and again, investment in agriculture was lower than had been anticipated, while the cost of farming was rising with the increase of the price of seeds, chemical fertilizers and pesticides. This brought about negative consequences for the development of agriculture. In the past few years, the major importance of developing agriculture has been reiterated.

It is also significant to stress the concept of macro agriculture. Long before China's economic reforms, the principle of "taking grain production as key" prevailed in agricultural production and had brought about serious consequences. For instance, in the pastoral areas of the northwest part of the country, vast expanse of grasslands were destroyed for the purpose of developing grain farming. This seriously damaged animal husbandry and failed to bring about good grain farming in the traditional pastoral areas. It was not until the principle of "taking grain production as key" was abandoned and the concept of macro agriculture was established and spread throughout the country that the mistake was corrected and the grasslands and animal husbandry gradually recovered. The development of macro agriculture and township industry had broken up the unitary formation of the rural economy and played an important role in the development of the rural economy as well as the improvement of the living standard of the peasants.

STRENGTHENING INFRASTRUCTURE AND BASIC INDUSTRY

With regard to strengthening infrastructure and basic industry, it is pointed out in the Outline that the serious lag in infrastructure and basic industry should be mitigated and infrastructure and basic industry should be developed to make it gradually conform with the overall economic development.

According to the Outline, the strengthening of railway transportation capability should be taken as the key to the development of transportation and the development of trunk lines become the central task. The superiority of a multi-channel system comprised of highway, river and air transportation should be maximized, to accelerate the construction of a comprehensive transportation system. For the development of telecommunications, the basic requirements should be high speed, high quality and large scale. International advanced technology and equipment should be adopted, the proportion of localization should be increased as quickly as possible and progress should be made to promote the construction of a high speed information network step by step.

According to the Outline, in the energy industry, it should be a principle to place equal emphasis on both the development and the conservation of energy. Coordinated development should be carried out among energy, economy and environment. A general guideline for the development of coal, petroleum and power industries was put forward in the Outline.

According to the Outline, with a view to accelerating the development of the construction of infrastructure and basic industry, a series of policies will be adopted by the State. The initiatives of both central and local governments should be maximized, and a clear-cut division of responsibilities will be worked out between the central and local government. A special plan for the construction of infrastructure and basic industry will be formulated to promote sound development. A long-term investment and financing system under policy considerations should be established and improved, to provide for the development of the construction projects encouraged by the government with financial support for policy considerations. It is necessary to encourage and direct funds from various sectors of society to invest in the construction of infrastructure and basic industry. When issuing stocks and bonds, priority should be given to the needs of constructing infrastructure and basic industry. It is also necessary to broaden the scale and scope of using foreign capital and to encourage foreign businesses to invest directly in infrastructure and basic industry. Price structuring should be readjusted further to maximize the important adjustment role of the pricing mechanism. The control of the pricing of products and services in monopolized sectors should be continuously strengthened. The policy of requisitioning land at low price for development of infrastructure and basic industry will be continued. Revenue obtained from the transfer of land resources by the government should be used to develop infrastructure. Investors in the construction of transportation facilities, through approval, may be permitted to have priority in obtaining the right to operate real estate development along the transportation lines and in the areas adjacent to the ports and airports as comprehensive economic compensation.

ACTIVELY REJUVENATING MAINSTAY INDUSTRIES

According to the Outline, the development of the machinery and electronic, petrochemical, automobile and construction industries should be accelerated, to build them into mainstay industries of the national economy.

The development of the machinery industry should include key basic machinery, basic parts and components and whole sets of equipment for major technologies as focal points, to optimize product mix and to improve the level of manufacturing technology and competitiveness. The development of the electronic industry should take micro-electronics as a base and the rising information industry, including telecommunications and computer, as the main body in order to accelerate the pace of modernization. For the development of the petrochemical industry, it is necessary to actively enlarge the production scale, upgrade technological levels and improve the depth of processing. For the automobile industry, it is necessary to form a production system with fewer production units, a larger volume of production, and a market composed of orderly competition, as quickly as possible, to increase domestic market share and improve international competitiveness. For the construction industry, it is necessary to focus on urban and rural home-building, to emphasize key State projects and city and township construction, to actively establish a unified, open, competitive and orderly construction market and to improve the quality of construction products.

It is pointed out in the Outline that the State will accelerate the development of mainstay industries through a series of measures. A unified industrial policy will be formulated and issued, and the implementation of the policy will be ensured through legislation and decrees. An investment and financing system in favor of the development of mainstay industries as well as a normalized direct financing mechanism will be set up gradually. Arrangements will be made to give priority to mainstay industries involved in annual stock and bond distribution. The government will support mainstay industries both financially and materially in their technology development in major fields and will, after approval of the State Council, provide a small number of large enterprise groups with the power to access overseas financing. According to international convention and relevant agreements, part of the mainstay industries' products may be viewed as immature industrial products and be given certain appropriate protection in a limited time frame. In the meantime, with a view to acquiring key technologies and equipment, it will be required to open up part of the domestic market as an exchange.

DEVELOPING FOREIGN TRADE IN AN ACTIVE WAY

It is pointed out in the Outline that foreign trade should continue to expand, trade composition be actively readjusted, the comparative superiority of China be maximized and efforts be made to improve export benefits. This is to help promote the readjustment and optimization of industrial formation, strengthen international competitiveness and keep the equilibrium of international payments.

The State will encourage the expansion of export of the following products: agricultural and side products with comparative advantages, and light industry and textile industry products; electronic durable appliances and other machinery and electronic products of which the manufacturing technology tends to be mature; products with high added value and international competitiveness and high and new technology products. The State shouldn't encourage the export of staple resource intensive products, which are in short supply domestically, and should gradually reduce the export of primary products and products with a high energy content. Some of these products may be restricted or prohibited from export.

The State will encourage the import of new technologies and key equipment and key parts and components related to these technologies; appropriately expand the import of primary products that are in short supply domestically; support immature industries to accelerate importation, assimilation and absorption of new technologies and production of new equipment and key parts and components. The State does not encourage the import of high level consumer goods.[2]

In readjusting foreign trade composition, the State will adopt the following measures:

- To maximize the important role of China import and Export Bank in supporting the increase of the export of equipment in whole sets as well as machinery and electronic products. On the management of the import and export products under quota restriction, China's Import and Export Bank will carry out quota bidding, auction or distribution according to rules based on the principle of efficiency, fairness and openness;

- To improve the policy for the promotion of the export of intensively processed, high added-value finished products and equipment in whole set;
- To offer foreign trade operation rights to enterprises under certain conditions;
- To encourage large scale enterprises (group) to set up overseas channels for direct sales; and
- To reasonably readjust the customs tariff rate according to the State industrial policy.

INDUSTRIAL ORGANIZATION, MANUFACTURING TECHNOLOGY AND INDUSTRIAL DISTRIBUTION

To improve the industrial organizational structure, the key is to promote the development of a rationally competitive market and realize economies of scale. The small industrial scale has been a serious barrier to promoting the growth of economic benefits and the international competitiveness of China's industry. However, the promotion of industrial economies of scale should take into consideration the specific characteristics of different industrial segments. Therefore, the Outline puts forward separate guidelines for the readjustment of industrial organizational structure for different industrial segments.

The Outline points out that the goal of the policy of industrial organization is to promote rational competition among various enterprises, to realize economies of scale, to coordinate different specialties, and to form an industrial organizational structure that conforms to the technical economic characteristics of different industries and the current stages of economic development in China. In respect to the industries with notable effect in a scale economy, it is necessary to form a market composition with large scale enterprises (groups) as main competitive bodies; for industries in which products are composed of a large number of parts and components, it is necessary to form a market composition comprised of large, medium-sized and small enterprises with both a rational division of work and coordination among one another at an appropriate scale; for industries without any notable effect in a scale economy, encouragement should be given to the development of small enterprises and the formation of a competitive market comprised of a large number of large, medium-sized and small enterprises.

In order to realize the readjustment of the industrial organizational structure along these guidelines, it is necessary to break down partitions between different departments and regions. Under the planned economy, enterprises were subordinated to various administrative departments under either the central or the local government. This inevitably brought about partitions among different central departments or different regions. Such partitions brought about the dispersion of industrial organizations, smallness of economic scale and barriers to the growth of unified market. Therefore, in an effort to promote the development of rational market competition and the growth of economies of scale through industrial organizational readjustment, the Outline puts forward a series of measures. For the industries with a natural regional monopoly, it is necessary to gradually introduce a market mechanism and encourage reasonable competition. For the industries and products with notable effects of economies of scale, the criteria for minimum economic scale will be formulated in succession. In the meantime, it is imperative to break up the divisions between different regions and different departments and to restrict or prohibit the development of projects not up to the criteria of economic scale, so as to promote the realization of scale economy. It is necessary to encourage enterprises, by way of fair competition, merging or mutual share holding, to autonomously carry out a combination and reorganization or form enterprise groups engaged in transregional, transdepartmental, transownership or even transnational operations. It is imperative to accelerate the formulation of decrees on market competition, in order to create a sound external environmental for fair competition among various enterprises as well as for the readjustment of industrial organizational structure.

The focal points of the manufacturing technology policy are to promote the development of applied research, encourage the integration of scientific research with production, accelerate the extension of research results, promote the introduction and assimilation of foreign advanced technology, improve the quality and technological performance of domestic products significantly, and reduce energy and material consumption as well as production cost by a big margin, so as to improve the technological level of China's industry.

In order to promote the technological progress of industry, it is necessary to adopt the following measures: Increase the investment in R&D from multiple channels and in diversified forms, so as to increase its proportion in the GNP; formulate and carry out key R&D programs of major importance to industrial development sector by sector and support and encourage the assimilation, absorption and development of imported advanced technology; strengthen the planning of the high and new technology industry and to effectively construct high and new technology development zones approved by the State; give impetus to the process of standardization and systematization and encourage the adoption of international standards, foreign advanced standards and stricter internal standard for enterprises; strengthen the new product development capability of enterprises through their own initiative and encourage enterprises to strengthen their ties with research institutions and higher education institutions, so as to accelerate the speed of the commercialization of research results; and publicize periodically, in the forms of decrees, which outmoded manufacturing technologies and equipment are to be eliminated.

The chief principles of industrial distribution are: When continuing to maximize the superiority of economic developed areas and to speed up the development of such areas, it is necessary to actively support the economic development of less developed areas and gradually narrow the gap between the economically developed areas and less developed areas; to support the development of an industrial belt, which may maximize the utilization of regional natural resources and economic superiority, and manifest the specialized division of work and coordination between different regions.

It is necessary to gradually form a rational layout of coastline areas, areas along Yangtze River and trunk railway lines and bordering areas and to take the large cities situated at trunk transportation lines as centers to further the development of large economic regions. In the eastern part of coastline areas, efforts should be made to develop an export-oriented economy, to take the industries and products with high added value, high hard currency earning rates, high technology contents and lower energy and raw material consumption as focal points of development, and to use more foreign capital and resources to achieve sustained and fast growth and better benefits. In the middle and western part, it is necessary to maximize its superiority in resources and its geographical superiority in bordering areas, which is advantageous to opening up to the outside world, and to strive to develop superior industries and products. The State will gradually transform from slanting to specific regions to slanting to specific industries in terms of economic policies, including investment, loans, projects distribution and the use of foreign funds, and give the necessary support to major projects in the development and construction of the middle and western part of the country. It is necessary to support the middle and western part to develop township industry and encourage economically developed areas to carry out joint development, technology cooperation, area to area support and personnel exchange with the middle and western part.

It is necessary to direct the healthy development of urbanization in its natural course of development and to form a city and township system with coordinated distribution of large, medium-sized and small cities.

The importance of industrial policy lies in its influence in the direction of investment, the input of material and human resources and the formulation of policies of encouragement and preference. The State is trying to realize the anticipated goal of the industrial upgrade through the aforesaid aspects. However, under the condition of a market economy, the government's manipulation can only be indirect; and inasmuch as there is an important factor of changing market demands, the outcome may not conform to the anticipated goal. For instance, the automobile industry has been one of the mainstay industries, but the growth of automobile industry hasn't been able to reach the anticipated speed. In the past few years, the emphasis on real estate development has tried to shift to home-building in the city, but so far as there is a large gap between the average purchasing power of city dwellers and the average price level of homes, the speed of the development of the construction of city home-building has slowed. In contrast, some consumer products industries have developed at a speed far beyond the original expectation. Nevertheless, the general direction indicated by the industrial policy will play an important role in long range industrial development and provide a framework for the technology policy to follow.

NOTES

1. In addition to the technology policies in 12 industrial sectors, the policies of information technology and biotechnology were declassified and published together with the technology policies of the 12 sectors in 1988.
2. See "Industrial Catalog Guiding Foreign Investment." The Catalog was first published in June 1995 and was approved by the State Council in December 1997 after revision and jointly promulgated by the State Planning Commission and the Ministry of Foreign Trade and Economic Cooperation. The first edition was published by *Beijing Review* in Issue 38, 1995, and the second edition was published by *Beijing Review* in March 1998.

11

The Development of Agricultural Science and Technology

The world acknowledges that China has successfully supported 21% of the world population with a total arable land accounting for only 7% of the world. Indeed, China has made accomplishments, since the founding of the People's Republic, in protecting, utilizing and exploiting the land. In the past 20 years, since the start of economic reform, great changes have taken place in China's agriculture. The change is manifested in four notable aspects.

The first striking change has been implementation of the output-related contract responsibility system. Land is a basic material of production in rural areas. The output-related contract responsibility system doesn't change the ownership of the land. The land is still owned by the State. What has changed is the forms of operation and distribution. That is to say, farming is no longer under the control of the local administrative unit. Previously, either production or distribution was under the administrative control of the People's Commune. The farming plan was issued by the Commune, and the distribution of the harvest was also under the direct control of the Commune. After the Great Leap Forward, the system of taking the production teams as an accounting unit was carried out.[1] It meant that the production plan and daily farming were under the management and control of production teams, and the harvest was also distributed under the control of production teams. Either way an egalitarian distribution system was carried out. This didn't bring about any incentive to farmers. Insofar as the farming plan and daily operation were controlled by the production team, farmers didn't have any initiative. Under the output-related contract responsibility system, though the land was still owned by the state, the right to use the land was distributed to farmers on a household basis. Now, the farmers may decide how to farm their own land, and they may reap their harvest after paying an agricultural tax. The meaning of "output related" is that the farmers should ensure a certain contracted level of output as their responsibility, as insurance for the overall harvest of crops. The new responsibility system realized a dream cherished by farmers for years, "Let farmers have their lands," and has greatly mobilized the initiative of farmers. The original duration for the contracted responsibility systems was 15 years. It has been extended for another 30 years.

The second important change is the cancellation of the unified purchasing and selling system. Previously, in the planned economy, all agricultural products were purchased by the State and sold by the unified distribution system of the State. Now, the pricing of all agricultural products is deregulated and adjusted by the market, except for cotton, tobacco and silkworm, which are still under State control. The pricing of main agricultural products, such as grain, eatable oil and meats, has been deregulated all over the country.

The third change was directing agricultural production to a macro-agriculture perspective. Previously, the guiding principle in agricultural production was taking grain production as the primary concern. This one-sided guiding principle had caused a series of losses, such as the destruction of forestry and grasslands in order to construct farmland. Now, through a macro-agriculture perspective, when making efforts to develop grain production, forestry, animal husbandry, fishery and various side productions are also developed. This is one reason for the faster development of agriculture.

The fourth change is the development of township industry.[2] A large number of farmers have been encouraged to enter circulation fields and transfer to non-agricultural industries. By the end of 1994, total revenue of the country's township industry was about 1,500 billion yuan and the total number of farmers working with township industry came up to 110 million. The rapid development of township industry did not rely on the investment by the State. The State had provided township industry with only partial support in bank loans and some preferential policies. Starting capital for township industry basically came from the accumulation of the collectives and farmers themselves. Now, township industry has developed into a powerful mainstay of the national economy. Township industry is the offspring of China's new commodity economy. Right from its birth, township industry has not required any preferential treatment from the planned economy, but has developed under the direction of a market economy and thus permeated with vigor.

The changes in China's rural area in the past two decades are profound and the achievements are great. Nevertheless, the general situation of long range agricultural production remains questionable. First of all, the basic condition in China is a large population with limited arable land. The aggregate territory of China is 9,600,000 square kilometers, about one-fifteenth of the world land mass. However, because of the large population, per capita land area is only one third that of the world. In addition, of the aggregate land area, deserts, glaciers, Gobi and stone mountains are unusable and make up a considerable part. According to statistics, there are 1.43 billion mu (one mu equivalent to 0.1647 acres) of arable land in China, 1.3 mu per capita, only one fourth of that of the world and one ninth that of the United States. The reserve resources of arable land in China are insufficient. According to statistics, existing wasteland is 500 million mu; however, only 170 million mu are suitable for opening as arable land. In accordance with these figures, though China is a country with vast territory, viewed from per capita occupancy, China's land resources are relatively poor. This should be the starting point for studying agriculture in China.

Second, China's population is currently rapidly expanding, while arable land is rapidly decreasing. This is a serious challenge. According to statistics, during the period of 1957 to 1986, the accumulated decrease in arable land was 610 million mu and the net decrease 230 million mu, for an average decrease of 790 thousand mu per year. After the issue and implementation of the "Law of Land Management of the People's Republic of China" in 1986, the trend of rapidly decreasing arable land was under control. However, in recent years, the trend is rising again. In the year 1993, the arable land of the country decreased 9.37 million mu, equivalent to the total area of arable land in the Qinghai Province. With population growth and the decrease in arable land, the situation called for serious attention.

A large part of the decrease in arable land in China is owed to some artificial reasons. For instance, in 1992, because of various kinds of construction, the internal structural readjustment of agriculture, and the establishment of development zones, as many as 10 million mu of arable land were occupied. During high speed economic development, as a result of the readjustment of the industrial structure and the acceleration of urbanization, generally speaking, it is inevitable to have occupancy of arable land. However, it is intolerable to have such a high occupancy rate in a country like China, which suffers from lower per capita land occupancy. In particular, a considerable part of the artificially occupied arable land belongs to illegal and reckless occupancy. Certain decrees were formulated and a series of administrative steps were taken to prevent further illegal and reckless occupancy of arable land.

Another serious land crisis in China comes from a lack of water, erosion, and the deterioration of the ecological environment. Now, the total area suffering from lack of water and erosion of soil has reached 1.3 million square meters, and the total area of desertification of land has risen to 176,000 square meters. A vast expanse of farmland is lost every year.

There are three options to tackle this difficult issue regarding arable land in China—the protection of arable land, the exploitation of reserved resources of arable land and tapping the potential of existing farmland.

TAPPING THE POTENTIAL OF EXISTING LAND

According to statistics, by the end of the 1980s, in existing arable lands, high-yield, middle-yield and low-yield farmland accounted for 24.9%, 34.1% and 41% of the total existing arable land, respectively. The existing arable land had great potential to increase output. Should farmland all be upgraded by one level, the grain output would increase notably. According to estimates, the potential of grain output from the existing high-yield, middle-yield and low-yield farmland was 24 billion kilograms, 55 billion kilograms and 65 billion kilograms, respectively.

To tap the production potential of existing arable lands, there should be overall disposition of agricultural development. Since the potential production of middle-yield and low-yield farmland accounts for 83.6% of the total, they should be taken as the focal point. To approach it from geographical distribution, the potential of the northern and eastern parts of China was the largest, accounting for half of total product potential. To approach it from degree of difficulty, the slope farmland in the south and the saline-alkaline soil of the north and northeast parts of the country were easier to exploit, while the poor dry land in the area that suffered from lack of water and the erosion of soil was more difficult to exploit.

There were quite a few conditions favorable to exploiting middle-yield and low-yield farmland, such as the availability of mature technology, flexibility in terms of scale and speed of exploitation, less investment relative to opening up wasteland and more channels for raising capital and faster effects. The investment in the use of middle-yield and low-yield farmland, according to estimates, by the end of the 1980s, was 15–40 yuan per mu on average, while the investment in opening up wasteland was about 200–500 yuan per mu on average.[3] In the past three decades, a vast expanse of low-yield farmland has been exploited and thus produced much experience.

For instance, the northern part of Jiangsu had been an area of repeated natural disasters and low yields. Because of the exploitations of middle-yield and low-yield farmland and the development of soil fertility on a large scale, grain output in the same area had sustained growth by a big margin. By the end of the 1980s; the unit output was close to that of the Taihu area, which had been a stable high-yield area of grain. By the late 1980s, output in the northern part of Jiangsu made up two-thirds of the total output of Jiangsu Province.

THE EXPANSION OF ARABLE LAND

Wasteland appropriate for farming includes wasteland with certain conditions (mainly water resources); after the land is reclaimed and cultivated, agricultural crops and economic forests can be grown or the land can be used for animal husbandry. According to statistics, the aforementioned reserved resources account for 190 million mu in China. In addition, open cut mines spread all over the country have occupied a land area of 30 million mu. To calculate based on a 50% reclamation rate, 15 million mu will be available for construction after reclamation and cultivation.

The characteristics of the farmland reserved resources in China are:

1. The total area of farmland reserved resources is smaller, accounting for only one tenth of the total existing farmland and one hundredth of the resources of wasteland appropriate for farming in the world.
2. The wasteland in China is unevenly distributed, and 54% of it is located in the northeast and northwest part of China, where productivity is lower.
3. The wasteland is of low quality. Nearly all wasteland suffers from restrictive factors such as lack of water and erosion of soil, desertification, salinization, drought, flood and infertile soil. It will be a heavy task to reclaim and improve the ecological environment and prevent the deterioration of the ecological environment.

Of the existing 190 million mu wasteland appropriate for farming in China, there are possibly 125 million mu available for the production of agricultural crops. These lands are mainly distributed in the northeast and northwest part of China, where the productivity of soil is low, productive conditions are worst and the consumption market is farthest away. The total production potential is about 25 billion kilograms. Key water conservation projects and other reclamation measures are needed for most of the soil. According to estimates by the late 1980s, investment would be as high as 200–350 yuan per mu. Fifty-seven million mu are appropriate for the development of fruitwood economy, distributed mainly in areas of hilly soil textured land in the south, with higher land productivity and economic benefits and facilities for raising capital.

THE PROTECTION OF FARMLAND

Farmland area is shrinking chiefly from the presence of capital construction and production and loss due to natural factors such as sandstorm, salinization, lack of water and the erosion of soil.

Since the founding of the People's Republic, up to the late 1980s and early 1990s, 470 million mu of farmland was occupied for production and living, 12.5 million mu per year on average. While it is inevitable for economic construction and development to occupy some farmland, in a large agricultural country with a population as large as 1.2 billion, with limited per capita farmland and reserved resources, the loss of 100 million mu of farmland every 10 years is a serious problem. In the coming years, the conflict between the occupation of farmland for the production and people's living and the development of agricultural production will be more intensified. The illegal occupancy and misuse, as well as the waste, of farmland should be strictly controlled. It is key for the protection of farmland.

The shrinkage of farmland is also derived from the deterioration of the ecological environment. In three decades before the end of the 1980s, the area of desertification in the agricultural and pastoral crisscross area in the northern part of China had increased to 50.6 million mu, an increase of more than 1.60 million mu per year on average, including 43% derived from overreclamation. In the existing farmland, there was about 50 million mu impacted by desertification and 290 million mu damaged by the lack of water and erosion of soil, accounting for 20% of the total farmland. Another threat is salinization.

In order to settle or to mitigate the threat of desertification in the existing farmland in the north, farming would have to be ended in 20 million mu. In the yellow soil, high plateau, hilly gully area, farming in 60 million mu would have to end. In the south, there were 20 million mu of farmland on the steep sloping fields with a slope of 30% that should be withdrawn from farming. Altogether, farming in 100 million mu should cease. Certainly, in order to protect the basic construction of the farmland and prevent the deterioration of the ecological environment, it is necessary for part of the land to end farming temporarily; however, in the long run, effective measures must be adopted to stop the swallowing of farmland by natural factors and to limit land loss by ending farming.

STRATEGY AND COUNTERMEASURES FOR FARMLAND CONSTRUCTION

According to the actual circumstances, the strategy for farmland construction in China is to protect the farmland resolutely, to make efforts to reclaim wasteland in a limited way and make the exploitation of middle-yield and low-yield farmland key.

In order to protect the farmland, it is imperative to prohibit the illegal use and control the misuse of farmland through legislation and regulation. In 1994, the State Council issued the "Regulations for the Protection of Basic Farmland," which stipulates the establishment and the protection of the basic farmland reserve area. According to the Regulations, the basic farmland reserve area includes (1) the production bases of grain, cotton and edible oil and agricultural products of famous brands and fine, special and new varieties; (2) high- and stable-yield fields and farmland with good water conservation as well as water and soil maintenance facilities or the middle-yield and low-yield fields that have been addressed and transformed or under transformation; (3) the vegetable production bases of large and medium-sized cities and (4) agricultural research and teaching experimental fields. Once the reserve area is designated, it shall not be changed or occupied by any units or individuals. In the event that it is unavoidable for State key construction projects of energy, transportation and water conservation to occupy part of the reserve area, the relevant authority should be applied to in accordance with the "Law of Land Management of People's Republic of China." Aside from taxes paid for the occupancy of farmland within reserve area by non-agricultural construction, the user shall open up an area of farmland as large as it occupied. In case the user is unable to open up the farmland on its own, it shall pay the land reclamation fee. The units or individuals who engage in agricultural production on the farmland within the reserved area should maintain and cultivate the land fertility. The State encourages the agricultural producers to use organic fertilizer, chemical fertilizer and pesticide on their farmland. The Regulations protect basic farmland with a legal basis.

Both the exploitation of middle-yield and low-yield fields and the reclamation of wasteland need substantial investment. One way to raise capital is to put up the funds collected from the occupancy of farmland for investment in using middle-yield and low-yield fields or the reclamation of wasteland. In addition to the initial charge on land occupancy, limited amounts of farmland construction fees may be charged annually. The measure can be used as a restriction on land occupancy and at the same time as a source for raising capital for farmland construction.

Nonetheless, the key is how to effectively exploit the middle-yield and low-yield fields. Aside from the formulation of necessary policies, it is imperative to further the application of agricultural science and technology. To develop agricultural production by relying on technological progress is one direction to pursue the potential of existing farmland.

In 1993, the "Agricultural Development Outline of China in the Nineties'' was issued. According to the Outline, the general direction of agricultural development in the 1990s is as follows.

Agricultural production should adhere to the principle of adapting measures to local conditions, to give full play to the superiority of local natural resources as well as the local economic and technological superiority. That is to say, if the local conditions are suited to agriculture, priority should be given to agricultural production; if the local conditions are suited to forestry, priority should be given to the development of forestry; and so on. Promoting the all-round development of agriculture, forestry, animal husbandry, side-production and fishery will bring about the regional, specialized and commercialized development of agricultural production.

According to the Outline, the State encourages and supports the development of grain production. The production and distribution of grain all over the country should adhere to the principle of taking each province (or district) as a self-sufficient unit with some surplus or as basically self-sufficient with a small part of its grain supplied by neighbors, to set up a reasonable layout. Various regions should develop measures to suit their own situations as they set up their commodity grain bases. It is necessary to adjust the structure of the varieties of grain according to market demands, to make an effort to develop grain of fine quality, and to increase the output of grain for forage use. In the middle and lower reaches of Yangtze River and the southeast coastline area, more fine quality rice should be grown, and in Yellow River and Huai River area as well as northeast China, it is necessary to actively develop the production of wheat, corn and soybeans.

The layout of economic crops and other crops should be further readjusted in accordance with the principle of suiting measures to local conditions, maximizing advantageous conditions and avoiding the disadvantageous and concentrating efforts in a proper way. The cotton production should be developed mainly in the Yellow River and Huai River area, the middle and lower reaches of Yangtze River plain and Xingjiang. With respect to edible oil production, rapeseed production in Yangtze River Valley and the south of Yangtze River and the peanut production north of Yangtze River should be the key industries, while giving consideration to the production and development of other edible oil species in various areas. For the production of sugar, sugarcane in the southeast coastline area and sugar beet in the northeast and northwest of China should be stably developed. At the same time, the development of the production of sugarcane in Yunnan and Guangxi should be accelerated.

Forestry production should proceed from the cultivation of forestry resources, the increase of forest vegetation and the strengthening of the supply of lumber and forestry products to adjust the layout reasonably. It is necessary to build up a number of key protection forests, water and soil maintenance forests, and farmland protection forests along the Yangtze River and coastline areas as well as in northwest, the northernmost part of China, the western side of the northeast, Taihang Mountain and the plain area, which have suffered windstorm and sandstorm damage and a serious lack of water and the erosion of soil. It is also necessary to construct a number of fast-growing, high-yield commercial timber forests by way of intensive operation concentrating on the east, central south and southwest parts of China. The bases of woody oil, perfume, medicinal materials and dry fruits should be constructed in the south of the Yangtze River where suitable conditions are available. In the shallow mountainous and hilly areas and the arid and semi-arid lands of the north part of China as well as in the coastline areas, it is necessary to develop fuel forests to relax the tension of energy resources in rural areas and promote the comprehensive utilization of agricultural residues and the development of methane.

With respect to the development of animal husbandry, in the suburbs of large and medium-sized cities, it is necessary to focus on the development of livestock, poultry, eggs and milk production, to meet the consumption demands of city inhabitants for fresh and live livestock and poultry products. In the rural area, it is necessary to continue to stress the production of pork swine, to enhance productivity, and to actively develop the production of lean meat swine. In the meantime, it is necessary to actively extend the ammoniated stalk and straw technology for the development of cattle rearing in rural areas. In the semi-agricultural and semi-pastoral area, it is necessary to maximize the superiority of the richness of forage resources, to accelerate the production of plant-eating livestock, such as cattle and sheep. In the pastoral area, it is necessary to pay attention to the construction of grasslands and the improvement, exploitation and utilization of meadows in addition to the prevention of the degeneration and desertification of the grasslands, in order to enhance the capability of meat production. It is also necessary to actively develop wool production, to meet the demands of the developing textile industry.

For aquatic production, it is necessary to develop sea and freshwater breeds, to actively exploit deep sea fishing, and to pay attention to the process and comprehensive utilization of aquatics. With respect to fishery, it is necessary to strictly control the intensity of inshore fishing and to actively exploit the fish resources in the middle and upper layers of the Donghai Sea and Huanghai Sea as well as those in Nanhai Sea. In the meantime, it is also necessary to accelerate the development of deep sea fishing.

Inasmuch as the overall layout of agricultural development in the 1990s had been mapped out, the important issue is realizing it. The realization of the overall layout should rely on the progress of agricultural science and technology. Only by relying on scientific and technological progress can the output per unit area of land and various agricultural resources be enhanced.

Specifically, in order to realize increased production in the 1990s the key is to grasp the enhancement of the output per unit area. It is imperative to rely on the progress of science and technology to improve the conditions of agricultural production and increase the output per unit area. In a decade, the grain and cotton output per mu should be increased 50 kilograms and 12 kilograms per mu respectively. In view of this, according to the Outline, in terms of agricultural science and technology, it is necessary to adopt the following measures.

It is necessary to organize the implementation of new water saving irrigation to enhance the utilization rate of irrigation water. It is necessary to focus on the different circumstances of various areas to make efforts in extending the water saving irrigation technology to develop water saving agriculture. It is necessary to extend technologies such as pipeline water delivery, channel lining, seepage control, spray irrigation and drip irrigation. The management and maintenance of water conservation facilities should be strengthened to improve efficiency. All measures are aimed at increasing the utilization rate of irrigation water from 40% to 50%.

It is essential to improve the levels of using fertilizer and pesticides. It is necessary to adopt the measures of applying fertilizer scientifically, such as applying deep fertilizer, applying fertilizer by prescription after surveying the soil and combining organic fertilizer and inorganic fertilizer, to increase the utilization rate of chemical fertilizer from 30% to 40%. Farmers should be trained to be able to use various kinds of pesticide in a timely, rational and scientific manner, to improve the level of comprehensive prevention and control of plant diseases and to eliminate pests. It is necessary to actively extend biopesticide and biological prevention and control technology, to reduce the losses of agriculture caused by plant diseases and pests.

It is necessary to adopt measures that comprehensively tackle problems and improve the quality of existing farmland. By transforming sloping fields to terrace fields, flattening land, improving soil, completing irrigation and drainage systems, cultivating land fertility and practicing scientific farming, one third of existing middle-yield and low-yield farmland will be transformed and will be able to increase output per unit area.

It is necessary to fully utilize resources of light, heat, water and soil, to improve farming technology, to increase the multiple crop index and to expand the sown area. Measures such as increasing the planting rate of winter vacant fields in the south, expanding the intercropping and interplanting areas and developing ratooning rice in certain areas in the south where resources of light and heat are insufficient for double rice but are sufficient for single rice should be taken, to make it possible for the multiple crop index of the fields all over the country to increase from about 150% at the end of the 1980s to about 160% by the end of the century. This would be an increase of more than 730,000 hectares of sown area per year.

Regarding the breeding, extension and dissemination of agricultural fine varieties, it is necessary to strengthen the agricultural basic and applied research, to store and breed a batch of high-yield, fine-quality and disease resistant and pest resistant crops, fine forest varieties and fine breeds of livestock, poultry and aquatics. In the meantime, it is also necessary to introduce a batch of fine varieties from abroad in a planned and selected way. When working to extend and disseminate various conventional fine varieties, the emphasis should be on the breeding of fine-quality and high-yield hybrid rice, hybrid corn and hybrid rape and on accelerating the dissemination of hybrid fine-quality varieties such as lean meat swine, cattle, sheep and poultry. The breeds used for the production of grain, cotton and edible oil should be replaced once or twice in the 1990s, and the dissemination rate of fine breeds of livestock, poultry, aquatics, sugar and forest fruit should be further increased. The appropriate agricultural technology should be disseminated with great effort. It is necessary to form a whole set of fine breeds and good methods, and to select a host of appropriate technologies with notable effects for dissemination. It is essential to increase the transformation rate of agricultural technology results from about 40% currently to 50% by the year 2000. It is necessary to continue to work to carry out the science and technology extension activities, such as the "Bumper Harvest Program"[3] and the "Spark Program"; to disseminate the agricultural technologies including pattern cultivation, dry farming, fields plastic film cover, optimized farming system, and a disease-pest-rat-weeds-calamities surveying and report and comprehensive prevention and control. It is necessary to extend optimized rearing technology for livestock and poultry, mixed feed and ammoniated stalk and straw technology, sea and freshwater high efficiency cultivation as well as the technology of resources increase and epidemic disease prevention and control. It is also necessary to extend and disseminate the technology of agricultural products preservation, process, storage and transportation.

It can be found from the Outline that the core of agricultural development in China is to increase agricultural output mainly by increasing the output per unit area. The increase in output per unit area must rely on the progress of agricultural science and technology. Emphasis has been placed on a few of the most important programs, and extensive efforts have been made.

One of these programs is the strengthening of construction of production bases for commodity grain, cotton, edible oil and sugar. By 1994, more than two hundred production bases for commodity grain were scattered over various counties all over the country. The plan is to construct more production bases for commodity grain and to bring the number of bases up to 500, and the comprehensive production capability of these production bases should rise to 50% of the total commodity grain for the whole country. All these production bases should be developed into agricultural bases with high productivity, fine quality and high efficiency. Likewise, a batch of production bases for cotton, edible oil and sugar should be constructed.

Another program works to accelerate the construction of a number of fast-growing, high-yield timber forests and economic forests.

A third program calls for strengthening the development of the "shopping basket" project. In China, the population of large and medium-sized cities is growing rapidly. For instance, the populations of both Shanghai and Beijing number over ten million. There are more than two hundred million people in urban cities. How to provide city inhabitants with sufficient quantities of vegetables, meats, eggs, milk and aquatic products at a price acceptable for ordinary people is an issue that not only is closely related to people's daily life, but also concerns the stability of society. The shopping basket project is aimed at promoting a satisfying daily life. Watching the housewife's shopping basket is an important measure to determine the standard of living of ordinary people. It is indeed a gigantic project to fill millions upon millions of housewives' baskets. In order to fill the baskets with meats, eggs, milk and vegetables, it is necessary to construct a large number of lean meat swine, beef cattle, milk cow, sheep and eggs and chicken commodity production bases. Modernized, intensive operation and management systems should be exercised and advanced technology should be employed for the development of these commodity production bases. Likewise, water resources of rivers, lakes and reservoirs should be utilized for the development of aquatic product cultivation bases. All these efforts serve to satisfy the housewives' shopping baskets.

While requiring more agricultural output from the soil, it is imperative to take care to protect, maintain and develop the soil's fertility. The construction of an agricultural support system is a long range program. To ignore the construction of a support system will inevitably bring about punishment from Nature and the society.

Water conservation is the number one issue in the construction of an agricultural support system. In a country with vast territory, various parts of China have different geographical characteristics and different needs for water conservation. The Three Gorges Project is mainly meant for the prevention of flood. However, North China suffers from a shortage of water resources. It is urgent to divert water from the south to the north, to address the problem of water supply in Beijing, Tianjin and other parts of North China. It is also necessary to complete a number of irrigation and drainage networks. For the maintenance of water and soil, while quickening the pace to remedy the lack of water and the erosion of soil in key areas such as Yellow River Valley and the upper reaches of Yangtze River, it is necessary to maintain the water and soil along the valley of Pearl River, Haihe River, Huaihe River and Liaohe River.

It is also essential to develop industries supporting agriculture. For instance, the development of industries of chemical fertilizer, pesticides, and plastic film and agricultural machines needs to be accelerated and geared to the needs of agricultural development.

The last and most important issue for agricultural development must be the development of agricultural science and technology and the improvement of the networks for dissemination and extension of agricultural technology.

The China Agricultural Research Academy, China Forestry Research Academy, China Water Conservation Research Academy and other research institutes engaging in the research of agricultural, forestry and water conservation sciences and technologies have been formed into a high level research system that provides advanced research results that are to be disseminated and extended throughout the country. These research institutions are required to strengthen basic research in bioengineering, including genetic engineering, cell engineering and fermentation engineering and, at the same time, to focus on the applied research of fine breed proliferation, cultivation technology, irrigation technology, water and soil maintenance technology, disease, pest and calamity prevention and control as well as the preservation, process, storage and transportation of agricultural products.

In order to bring agricultural scientific and technological results to the vast countryside, a series of intermediate links is needed. Therefore, aside from the encouragement and promotion of the integration of agricultural science and technology with production, it is necessary to develop a broad network to promote agricultural science and technology.

Since the founding of the People's Republic, agricultural technology extension work has experienced quite a few fluctuations. In the early 1950s China had built up an agricultural technology extension system with the participation of 95,000 full time technicians all over the country. However, during the Great Leap Forward, the system suffered serious frustrations. In the early 1960s, the agricultural technology extension work was readjusted and strengthened. Emphasis was on how to bring the broad masses of the peasants to farm scientifically through experiment, demonstration, farming on sample fields and training a technical force among the peasants. This effort of agricultural technology extension was seriously sabotaged during the period of the Cultural Revolution.

After the disintegration of the People's Commune, the government decided to restore and strengthen agricultural technology extension setups at various levels. Gradually an agricultural technology extension system formed by the county agricultural technology extension stations, township agricultural technology stations and village agricultural technology service groups was established. However, when the country started to transform from a planned economy to market economy, a number of places tried to push the agricultural technology extension stations onto the market and cut off their financial support. This seriously weakened the role played by these technology extension organizations.

Nonetheless, with the experiences of the past fluctuations of technology extension organizations, quite a number of local governments recognized the importance of agricultural technology extension networks, and a consensus was reached to strengthen the networks. It started with the action of the fixation of the nature, staff and appointments of the technology extension station (it was called "three fixations" for short). Since previously a number of local governments tried to have technology extension stations be run as businesses, it was necessary to fix the nature of the technology extension station to be a public service undertaking. The second fixation was setting up the size of the staff, and the third was deciding the specific personnel working with the station. Actually, with the "three fixations," the financial support of the agricultural technology extension stations was also fixed and thus ensured its normal operation. Jiangsu Province started the "three fixations" as early as 1991. By 1996, 23,188 local technicians had been engaged as staff of the agricultural technology extension stations all over the province. Anhui Province had a total staff of 11,000 local technicians working for the agricultural technology extension stations throughout the province.

The broad networks formed by the local agricultural technology extension stations has greatly promoted agricultural production, helped improve the quality of farmers and invigorated the rural economy. In 1995, the agricultural technology extension networks of Gansu Province carried out 470 large scale demonstration and extension projects to increase agricultural output. Among these, the cornfields plastic film cover technology had been carried out on more than 2.5 million mu of corn fields in 43 counties of nine prefectures and achieved an increase of 158 kilograms of corn per mu and a total increase of 397 million kilograms of corn. The wheat plastic film cover bunch planting technology was carried out on 10.8 thousand mu of wheat fields and achieved an increase of 70–100 kilograms per mu. In Gansu Province, 7,045 technicians and more than 12 million peasants had received training. Right now, there are 66,834 part time technicians of farmers at the township and village levels and 417,000 technology demonstration households all over the province.

A diversified agricultural technology extension system has now emerged. The system constitutes three categories of agricultural technology extension organizations. The first category is government undertakings, such as government agricultural technology extension departments and agricultural technology extension stations at the township level, which are the main force of agricultural technology extension. As government undertakings, these organizations offer free services for farmers. The second category is the operational type of technology extension service organization. It is mainly a service method of "company + farmers' household." The company provides farmers with seeds and technology, and the farmers are only responsible for production. The company then purchases the products from the farmers, processes them, and sells them at market. This is a way to link petty production with a big socialized market. The third category is a self-service type. The rural area Specialized Technology Society first emerged as an organization of volunteer farmers. By the late 1980s, such self-service organizations were widespread throughout the country. Now, through the societies' efforts, a large number of farmers are organized. The societies provide farmers with fine varieties and advanced technology, give direction for production and help purchase and sell products of the farmers. The societies provide a series of services. The diversified agricultural technology extension system as a newly emerging network has played an important role in promoting the development of agricultural production through the input of agricultural science and technology.

With a view to up scaling the dissemination and extension of agricultural technology, a few large scale technology extension programs have been developed. The "Bumper Harvest Program" is the one mainly serving for agricultural development. The Bumper Harvest Program was co-sponsored by the Ministry of Agriculture, Animal Husbandry and Fishery and the Ministry of Finance. It is a long-term program that started in 1987. The Program is aimed at accelerating the dissemination and extension of the research results and advanced technology in the fields of agriculture, animal husbandry and fishery. This is to promote the development of production of agriculture, animal husbandry and fishery and to realize the goal of higher output, fine quality, lower consumption and high efficiency. The purpose of the Bumper Harvest Program is to accelerate the promotion of advanced appropriate technology to enhance the per unit area output. The implementation of the Bumper Harvest Program has shortened the cycle of promoting advanced appropriate agricultural technology and broadened the scope of extension. On the other hand, it has changed the previous way of extending a single technology, and, instead, carries out the extension of comprehensive high-yield technology (i.e., to put single measures or single technology results together to form a complete set of technology) to achieve a more outstanding comprehensive effect.

The Bumper Harvest Program promoted cooperation among three agricultural systems, namely, the agricultural research system, agricultural education system and agricultural extension system. The Program is organized by the agricultural administrative department. So far as there are explicit targets, concrete measures and fund insurance, it is easier to organize cooperation among the three systems and to make the scientific and technological forces of the three systems fully display their superiority.

The contents of the Program center around key issues of agricultural development in the country. The main contents of the Program are as follows:

1. The extension of fine varieties;
2. The extension of high-yield cultivation technology model, the soil improvement technology applied to low-yield fields and various single output increased technology;
3. The extension of technology for the utilization of agricultural facilities, covered film fields and other chemical materials;
4. The extension of optimizing prescription fertilizer application technology and other scientific fertilizer application technologies;
5. The extension of energy and water saving implements and scientific irrigation technology;
6. The extension of technology for prevention and control of agricultural disease, pests, weeds, rodent calamities and epidemic diseases of livestock, poultry and fish;
7. The extension of optimized prescription feedstuff and scientific rearing and reproduction technology for livestock, poultry and fish;
8. The extension of concentrated cultivation technology of sea water and freshwater aquatics and proliferation technology of aquatic resources in large and medium areas offshore, as well as lakes and rivers;
9. The extension of the technology of preservation, process, storage and transportation of agricultural products;
10. The extension of appropriate advanced machine technology.

From the first year of implementing the Program, a number of accomplishments have been achieved. In the first year, 106 contracts were carried out, and the total planting area of various agricultural crops participating in the Bumper Harvest Program came to 42.05 million mu. The Program had provided 8.683 million commodity livestock and poultry and 2.46 million mu additional aquatic cultivation area. An increase of 1,913 million kilograms of grain output was achieved, accounting for 18% of the total increase of grain output in the same year. The increase of ginned cotton rose to 41.144 million kilograms, rubber 740,000 kilograms, milk 559,000 kilograms, livestock, poultry, meats and eggs 33.828 million kilograms, freshwater crab 770 thousand kilograms and fish 13.06 million kilograms. With 45 contracts carried over to the subsequent year, the total economic benefits of the Program came up to 1.6 billion yuan. The Bumper Harvest Program did make a breakthrough in making agricultural science and technology grow into agricultural production.

With limited per capita land resources, the development of agricultural production in China relied mainly on the enhancement of output per unit area and, therefore, has to develop intensive farming to claim more output from limited land area. The intensive farming requires more input, both material and science and technology input. In many areas of China, after implementing the contract responsibility system, the fields have been divided into small plots to be farmed by farmers' households separately, and multiple crops are planted on a single piece of farmland. This is unfavorable for the extension of specialized agricultural technology and scale economy, and especially unfavorable for the large scale extension and application of energy saving irrigation technology, plant protection technology and the mechanization of deep chemical fertilizer application. Therefore, the issue of how to develop economies of scale in agricultural production has been put on the agenda.

As the change from a contract responsibility system to a scale operation is a big change, the transformation will be a relatively long process. Currently, the scale operation is still in the initial stage. However, in different parts of the country, the scale operation has been carried out in different forms and has achieved some success.

The transformation from contract responsibility system to scale operation is a very sensitive issue, as it relates to the land use rights of farmers. Developing a scale operation is necessary in order to develop and modernize big agriculture. It is necessary to develop a rational flow system for land that had been contracted to individual farmers. However, the flow of land should never impact the land use rights of the farmers, as land is the basic production material for farmers or, in other words, the lifeblood of farmers. Therefore, the transfer of land should be treated as a commercial deal. As a commercial deal, it has to be on a voluntary basis. For instance, in Chuzhou of Anhui Province, there is a township where the farmers had planted rush for more than 20 years. However, as the rush planting was scattered over different plots of land for a long time, it was difficult to rotate crops, to guarantee quality and to make sure of the output. In order to develop rush production, the township set up a rush company in 1991, selected three villages with the convenience of transportation and better water conservation facilities and carried out a "lease back and contract back" experiment on a field of 2,000 mu. That is to say, on a voluntary basis, the company leased the farmland, which had been contracted to individual farmers previously at the rent of 180 yuan per mu, and then contracted the land back to those farmers who had rush planting experience and paid them 490 yuan per mu. The farmers who contracted the land should, according to the requirements of the company, select seeds, cultivate buds, manage fields, harvest, sell the crops and settle accounts in a unified way (and possibly receive a bonus for extra harvest). In some other places, other models of leasing are adopted to promote scale operations. In some places, in the fields planted high-yield crops, the village might rent back the land from individual farmers who had contracted the land. The original contracted farmers might have priority in contracting again for the piece of land; otherwise, the land might be contracted to a new contractor and the new contractor might be charged 160–200 yuan per mu and pay 140 yuan back to the original contractor. Through such arrangements, scale operation and intensive operation were carried out.

In various forms of scale operations, the farmer's benefits are addressed by the guarantee of the land use rights of farmers gained from the contract responsibility system. There is a principle to follow for various forms of scale operation. Instead of getting rid of the contract responsibility system, it is imperative to use the contract responsibility system as a basis for developing various forms of scale operation.

The development of agricultural science and technology is inseparable from the development of operating systems in rural area. The disintegration of the People's Commune and the development of the output related contract responsibility system had greatly stimulated enthusiasm and initiatives of farmers, as they were able to farm on land that they possessed the right to use. With the enhancement of the initiatives of farmers, agricultural science and technology was able to extend to their fields, which paved a way for the integration of agricultural science and technology with production. However, small scale production limited the material and technology input into the farmland and made it impossible to carry out intensive operations on the farmland. Furthermore, it is difficult for small scale and scattered agricultural production to conform with the changing demands of the market. The scale operation and intensive operation are necessary for the further development of agriculture. The scale operation and intensive operation will pave the way for further application and extension of advanced agricultural science and technology.

NOTES

1. During the late 1950s, when the People's Commune was first established, the Commune was the basic unit for accounting. All the harvest, after the contribution of crops to the State and the deduction of the part reserved by the Commune, was distributed to the Commune members according to the working

days of each member. As this system failed to give incentive to farmers, in the early 1960s it was changed to using the Productive Team as a basic unit of accounting, which meant to distribute the harvest in a smaller scope.

2. Township industries were developed by farmers in the villages. Since quite a number of these industries were located in small towns that were the hubs of surrounding villages, such industries were called "industries in villages and towns," translated into English as "township industry."

3. The figures here haven't been converted into the current value of RMB.

12

Township Industry and the Development of Appropriate Technology: The Spark Program

Township industry first emerged in the late 1970s and early 1980s. It was developed in rural areas by local farmers, who have been the only source of labor for township industry. China has a large rural population, and as farmland is limited, there is surplus of labor power. That's the reason that, once it emerged, township industry developed so rapidly. In 1984, the total output value of township industry was more than 100 billion yuan, by 1991, it had risen to more than 1,000 billion yuan, ten times of that of 1984. In 1991, the total output value of township industry was equivalent to that of State-owned industry. By 1994, the total output value of township industry came to one-third of the GNP in China. Right now, township industry is no longer a supplement of State-owned industry, but a vital new force in China's economic construction.

Township industry has played an important part in economic growth in rural areas and has helped to increase the income of peasants. In 1993, the total output value of township industry accounted for 71% of rural areas. Township industry has helped to settle the surplus labor power in rural areas. It is now able to absorb about 110 million of the surplus labor power of rural areas. Furthermore, township industry has broken a new path for the industrialization of rural areas (Table 12). In China, it is impossible for a large number of the rural population to move into large and medium-sized cities. The rapid growth of township industry has virtually turned a large number of rural towns into small urban areas

Table 12
General Situation of Township Industry by the End of 1995

Total Number of Township Industries	6.518 million
Total Number of Workers and Staff	73.005 million
Total Industrial Output in 1995	3,893.35 billion yuan
Average Growth Rate	29.7%

Source: China's Third Nationwide General Industrial Survey by the State Statistical Bureau.

that attract many of the rural population to work and live. There are about 20,000 such townships in the country right now. It is a new type of integration of rural and urban area.

As the development of township industry was spontaneous, in different areas there are different models of development. Usually, people divide the development of township industry into three different types: the Pearl River Delta type, the Southern Jiangsu type and the Wenzhou type.

All three types of township industry share some common characteristics: a large population and little farmlands; it has grown into the mainstay of the economy; the major part of the local labor force has transferred into non-agricultural industry and a large segment of the external labor force has been absorbed into these areas; the per capita income of peasants has increased substantially and is close to or has attained a comparatively well-off status; industrial formation has been improved, and the situation of a high-speed and coordinated development of first, second and tertiary industries is starting to appear; small towns are prospering; many township industries are gradually overcoming various kinds of problems which arose during the initial stage of development; the export-oriented economy has developed rapidly; and many township industries are expanding their operational scale and upgrading their technological level. Nonetheless, apparently, the three different types have a number of their own characteristics.

1. *Pearl River Delta.* The largest advantage of Pearl River Delta is its proximity to Hong Kong and Macao. With a large number of overseas Chinese, the policy environment of this area is comparatively relaxed. Shengzhen and Zhuhai in this area are both examples of reform and opening up to the outside world; the development of a commodity economy has been earlier and faster in this area. During the late 1980s and early 1990s, when township industries were readjusted nationwide and a large number of township industries suffered setbacks in other parts of the country, those located in this area had developed a large number

of export-oriented industries and opened to new prospects. The Pearl River Delta and Hong Kong mutually supplement each other's advantages and have mutually penetrated and merged into one entity, which has made market-oriented reform and the development of market economy in this area irreversible.

2. *Southern Jiangsu.* Southern Jiangsu is a richly endowed land. In Southern Jiangsu, the economic basis of the rural area was relatively stronger, education was comparatively well-developed and people's cultural and technological quality was higher. Township industry in this area had an earlier start. With support from Shanghai, Suzhou, Wuxi and Changzhou, township industry in this area has experienced smooth development. In the early stage of development, the township industries took advantage of their small size and operation flexibility. Inasmuch as they were small, they were flexible in changing their product lines to meet changing market demands. Now, the situation is quite different. Many township industries in this area are developing toward modern industry. Many of them are improving the quality and varieties of their products as well as efficiency. They have developed a number of joint venture enterprises and entered international markets. A number of township industries are developing high-tech products, to have reserve strength in domestic and international market competition. A momentum of mergers and conglomerations has appeared in a number of township industries in Jiangsu Province. More than 6,000 backbone enterprises have been formed. The total output value of these enterprises accounts for over 55% of all the township industries of the province. Two hundred of them have risen to a size similar to State-owned large and medium-sized enterprises, and one hundred of them have entered the high-tech field. Because of their cross linkages and combination with State-owned large and medium-sized enterprises, universities and research institutes, township industries have been imbued with new vitality. The number of joint venture enterprises developed by township industries in Jiangsu is about one thousand. It is noteworthy that, in 1991, there was a batch of townships, each with less than a hundred thousand population, taking township industries as their mainstay and having all-round development in agriculture, commerce, industry, housing, transportation and service. They had an annual output of around a billion yuan per township. This included the Shenze township of Jiangsu, which had an annual output of over 1.5 billion yuan. There were also villages with less than 1,000 population, each with an annual output of 200 million yuan. In 1984, there were six townships in Jiangsu, each with a total industrial and agricultural output of more than 100 million yuan. Eight years later, such townships increased to 550.

3. *Wenzhou.* In Wenzhou, 70% of the territory is mountainous area, and the area of flatland is very narrow. As Wenzhou has a population

of 6.8 million, the per capita area of farmland is little more than 0.4 mu. In addition, because of its special geographical conditions, it has received very little government investment since the founding of the People's Republic. Its collective economic base was weak and transportation conditions were poor. Wenzhou as a whole had been a poor area. Therefore, the rapid development of township industry in Wenzhou is not due to any natural conditions, but comes from artificial reasons, which mainly include (1) a variety of small commodities produced by farmers' family industry scattered over the whole area of Wenzhou; (2) a specialized marketing force composed of thousands upon thousands of salesmen; and (3) a batch of specialized fairs, which have been highly reputed nationwide.

Many successful businessmen in Wenzhou started with the production of a small commodity for daily use. Even the production of a small commodity requires an elaborate division of work. For instance, a button is a small commodity, but to produce a button requires people to source raw materials, people to mold, people to punch holes and people to polish. In addition to processing, it also needs a number of people to market the products generally as well as market to a specialized button wholesale market located in Wenzhou. This is a socialized division of work centered around the production and selling of buttons. In Wenzhou, for quite a while during the process of the development of township industry, each family took care of a specific link of the division of work, and the local family relationship served as the tie that brought families together to produce and market buttons. For people in Wenzhou, they didn't need a big factory to produce the commodity. The small township itself was a big factory, each family a workshop, and the family relationship brought various workshops together. People in Wenzhou were sure of how to make good use of human relationships, which was an important social resource. By utilizing such resources, they were able to turn a small township into a big factory. The trust between kinfolk and friends helped reduce transaction costs, while the division of work and cooperation among neighborhood families kept starting capital to a minimum and saved management costs. A social network formed by traditional ties of blood and geography thus grew to invisible capital and played the double roles described by economists as "human pool" and "capital pool." It might be the secret of people in Wenzhou, who were able "to start the economy from nothing." It was highly regarded by the experts as "breaking a path from the tradition to modernization."

People in Wenzhou also had a tradition of marketing. Many of them used to leave their hometown and go to remote places to do business. Now, even in Paris, there is a small Wenzhou part of town with a concentration of restaurants and small stores run by people a long way away from their hometown in China. The small garment manufacturers in

Wenzhou from time to time receive updated Paris fashion design information and drawings sent from Paris by fax by their kinfolk. The small merchants from Wenzhou, scattered over all parts of the country and world, not only did a good job in selling small commodities from Wenzhou but also collected information pertaining to market demand from every corner and fed it back to their hometown. In addition, in Wenzhou there are a number of large specialized wholesale markets with a variety of small commodities ready for selection by buyers from all over the country. There are more than ten thousand buttons in different patterns and colors on the button market. Aside from buttons, low pressure electric devices are also important "fist products" (main brand name products) of Wenzhou. There are 505 various commodity markets and 354 specialized markets in Wenzhou, including 16 that have an annual sales of more than 100 million yuan each. Each specialized market has its own "fist product."

Coming into the 1990s, people in Wenzhou put forward a new goal—"second pioneering." Now, the family industries that were the basis for the Wenzhou model, through various forms such as combined production, combined operation, joint venture and share holding, have gradually transformed into joint stock companies. The total number of joint stock companies is 36,887, and their total output value accounts for 80% of the entire township industry in Wenzhou. Another new development of the Wenzhou model is the integration of expanding the specialized market with the construction of new towns. A batch of new farmers' towns have emerged. The towns in Wenzhou have increased from 22 in 1978 to 134 in 1993. The industrial output of all these towns accounts for 80% of the whole area of Wenzhou, including nine central towns, each with a population of more than 60 thousand and an output value of more than 500 million yuan. It is in these central towns that the main products of Wenzhou (each has total sales over 100 Million yuan) are manufactured and 192 major joint stock companies (each with an output value over 10 million yuan) are located. The integration of central towns and specialized markets has promoted the concentration of a productive force and reduced the production costs and transaction costs of manufacturers, improved productivity and promoted industrial formation in rural areas as well as upgrading the level of industrialization as a whole.

Due to the rapid development of the township industry, from 1978 to 1993, the GNP of Wenzhou had increased from 1.331 billion yuan to 18.8 billion yuan, the total industrial and agricultural output value from 1.889 billion yuan to 39.489 billion yuan and peasants' per capita income from 113 yuan per year to 1,474 yuan per year.

The development of township industry has been by no means even and smooth, but along with disputes, and in the early stage of development, there were indeed quite a number of problems. Most township industries started with low technology and outdated or even primitive production conditions and were engaged in the production of labor intensive products. As a result, such industries couldn't help but have low productivity and high energy and material consumption. People of industrial circles kept complaining of the scrambling of energy and materials by the township industries with large and medium-sized industries. In addition, a number of township industries were the source of serious environmental pollution. For some time, a number of people objected to further developing township industry.

Nonetheless, once township industry emerged, it had a vitality that made it impossible to replace. China has a rural population of more than 800 million. Agriculture alone cannot bring about a relatively high income for peasants or enough jobs for the rural labor force. The large and medium-sized cities in China are unable to absorb and "digest" the millions upon millions of surplus labor force from the rural areas. These problems can only be solved by the development of township industries. Township industry doesn't need the State to make any significant investment beyond the financial power of the government. According to the statistics, in 1984 it required the increase of fixed capital investment of 14,000 yuan in State-owned industry to create a new job. As at that time, the labor force working with township industry numbered 50 million, a total investment of 700 billion yuan would be needed to create new jobs for the labor force working with township industry. Township industry was started and developed by peasants. They began with very limited starting capital (which they raised themselves) and developed it by relying on their abundant human and social resources. With traditional diligence and wisdom, peasants in China developed their own industry.

It was found later on that township industries developed by peasants not only helped the peasants but also supported the remolding of old large cities. For instance, old large cities such as Shanghai and Tianjin had an overconcentration of industry and serious environmental pollution and were in need of remolding. One of the major measures for remolding was to move out part of the conventional industries and develop township industries to make the large city a technology intensive industrial city. In the meantime, a number of large industries in large and medium-sized cities used township industries as suppliers of parts and components. For instance, there were quite a few brand-name consumer electronic products manufactured in Suzhou with 80–90% of their parts and components processed by township industries. Therefore, during the early stage of development of township industry, even though there were a number of problems, the township industry was indispensable. At the same time, it was figured out that many of the problems that arose during the early stage of development of township industries were derived from a lack of science and technology. Instead of getting rid of township industries, it was most important to help equip them with science and technology.

Of course, many township industries themselves were quite sure their technical force was insufficient, and many of them tried to hire full time or part time engineers to help them. Nonetheless, it was necessary to have a systematic input of science and technology in a planned and well-organized way. Under the circumstances, the State Science and Technology Commission developed the Spark Program. The Spark Program, as declared, served to rejuvenate the rural economy by extending science and technology, but during the early stage of development of the Spark Program, 80% of its efforts were focused on the extension of science and technology to township industry.

The Spark Program started in late 1985. At that time, it was recognized that only by relying fully on scientific and technological progress could small township industries be able to improve the quality of their products, reduce their consumption and enhance their productivity. Should the backward areas wish to enhance their competitiveness and develop a market, they needed even more to rely on scientific and technological progress. The prevention and control of environmental pollution also needed to rely on scientific and technological progress. Therefore, the development of the Spark Program should be a long range basic policy and strategic measure.

In order to fit the Spark Program to the development of local township industries, a practical goal was set up for the implementation of the Program: to constantly provide a large number of local and township industries with new designs, new techniques, new equipment and scientific management methods to help them, under the premise of constant improvement of economic benefit, to gradually transform onto the track of being "small, specialized, modernized and combined."[1]

In line with this practical goal, the guide for the selection of technology was to make it appropriate. The basic purpose of the Spark Program was to help local township industry to rely on scientific and technological progress to improve and gain economic benefits, social benefits and ecological benefits. Making it appropriate was most important for the practical development of these industries. In order to keep it appropriate, it was necessary to compare and select not only according to the advancement and level of the automation of technology but also according to market study and analysis of the possible benefits of adopting new technology in terms of productive conditions and raw material supply. It also needed to consider the capability of workers in mastering new technology. Furthermore, it was imperative to properly evaluate the ratio of performance and price, the ratio of cost and benefit and the market competitiveness of products. Only by maintaining "appropriateness" could progress be really "short, even and fast." "Short, even and fast" was originally a phrase used in volleyball. In volleyball, only by passing the ball with a short, even curve, as fast as possible, could a decisive stroke be given to the rival. In technology development, "short, even and fast" meant that only if the cycle of the commercialization of technology result were short could benefits and payoff be gained quickly; even the technology wasn't that advanced, but only practical and functional. Of course, technology selected under the aforesaid premise should be as advanced as possible. The technological strategy of the Spark Program could be described as "short, even and fast."

The technological strategy of the Spark Program was specified, and a number of projects were listed in the implementation plan. For instance, the implementation plan for the period between 1986 and 1987 included 736 projects, which were classified into 10 categories:

1. New methods of raising livestock and poultry and the development of a whole set of technologies for the storage, transportation, preservation of the products of livestock and poultry raising and local processing were to be introduced. The transformation of the traditional family livestock and poultry raising to large scale, new commercial feeding and management was an issue of major importance in promoting reform of the industrial structure in rural areas. It didn't need substantial investment and was easy to start up and easy to handle. It was able to develop value-added intensively processed products from grain and to promote the development of planting; it was able to increase the production of animal protein to improve the diet of urban and rural people; it was able to provide raw materials, such as hide, hair, feathers, meat, blood and bone, for secondary processing to prepare for the development of a series of new industries, such as storage, transportation and processing.

2. The comprehensive utilization, processing, and storage, transportation and preservation technologies of agricultural, forestry and native products were begun. The storage, transportation and preservation of agricultural and forestry products served for linking up the countryside with cities and should form a whole set of technologies to serve as a chain. The local growth process, storage, transportation and preservation should be the focus of the development of rural industry, especially for areas far away from large industrial cities. The development of these technologies are important for the formation of large scale planting commodity bases.

3. The construction of villages and towns, the development of building materials, and the promotion of the standardized, factorial and commercialized construction of houses for peasants are necessary. With the reform of the economic structure in rural areas, the peasants' income had increased notably, and peasants were eager to improve their living con-

ditions. During the sixth five year plan period (1981–1985), the total construction area of new buildings in rural areas rose to 3.8 billion square meters. This exceeded the total construction area for the previous 30 years. In 1985, the total expense for the construction of peasants' houses and public facilities came up to 45 billion yuan. Therefore, it was necessary to actively improve the technology of building materials and to promote standardized, factorial and commercialized construction of peasants' houses.

The development of new building materials was a pressing issue in construction. Solid brick made of loess and clay had been the traditional building material in China for thousands of years. In recent years, overconsumption of loess and clay by large scale construction has posed a serious threat to preserving farmland soil.

4. The mining of small minerals and primary processes and the development of mineral products should be undertaken under certain conditions. In the 1980s, there was a rapid development of small minerals, but the mining of small minerals continued to use traditional local methods that had high labor intensity and low return rates, a heavy waste of resources and serious environmental pollution. Some of the primary process products were of poor quality, and economic benefits were very low. It was necessary to adopt certain measures to make the exploitation of small minerals more cost effective and environmentally responsible.

In accordance with the circumstances of resources and market demands, the Program planned to focus on developing a variety of "small non-ferrous metals" such as tin, gold, copper, aluminum and antimony, to supplement "big nonferrous metals." Among non-metal minerals, it planned to develop kaolin, bentonite and olivine sand with high benefits and fast efficiency.

5. The extension and application of new technology and new materials were planned. In the 1980s, internationally, biotechnology, microelectronics technology and new materials were applied extensively in more fields than ever. Therefore, the plan was to select a number of appropriate new technologies that were mature, required less investment, and provided fast effects, and to use them to remold and equip small township industries. It was aimed, in this way, to upgrade their technology starting point and technological level, improve their productivity and the quality of products and increase their economic benefits. Taking biotechnology as an example, the main contents of technology development included (1) the manufacturing technology of new biological products, including edible and medical syrup, enzymatic preparations, diagnostic reagents, and nucleotides; (2) using new biotechnology to remold the traditional fermentation industry, including using solidifying enzymes and solidifying cells to produce alcohol, beer, salt sauce, vinegar and lactic acid; (3) embryo implantation technology; (4) plant tis-

sue culture technology, including the factorial cultivation of ginseng, cherry, banana and flowers; and (5) edible mushroom production.

6. Township industries should serve as a source of supply for large industries. For the time being, a number of small and township industries at various places were centered around the demands of large industries. These small and township industries processed parts, components and elements, provided raw materials and semifinished products for large industries, and utilized the products or side products of large industries as raw materials to carry out deep process and comprehensive utilization. In this way, they made themselves much stronger and, at the same time, created a new favorable environment for the development of large industries. The match of township industries with large industries should be an important part of industrial formation in China.

7. Regional comprehensive development was called for. In China, the mountainous area that makes up two-thirds of the total territory of the country has an abundant reserve of resources and also serves as an ecological protective screen for the plains area. For historical reasons, the level of economic development of the mountainous area was much lower than that of the plains area. It was an urgent task to change the seriously backward state of the mountainous area. Also, China has a coastline as long as 18,000 kilometers, with a broad area of mudflats and enormous potential for the development of multiple operations, which are of major significance for the provinces and districts with less land and high density population.

The Spark Program planned to work out an optimized program for all-round development and focus on a few mountainous areas. As for the development of the mudflats, eight sites had been selected as a demonstration program.

8. The Program planned to promote the establishment of commodity production bases and the development of a commodity economy and to promote the development of export commodity production bases mainly formed by small township industries. Such bases should be established in areas with a good foundation, a stronger capability of development and the convenience of transportation, such as the southern part of Jiangsu, places along Shanghai-Hangzhou-Ningbo as well as the provinces where the implementation of special policies for opening up was under way, such as Guangdong and Fujian. In order to mitigate the shortage of non-staple food supply in large cities, it planned to set up a number of commodity production bases of non-staple food stuff such as vegetables, aquatics, fruits, poultry and eggs and meat. In the north, it planned to set up bases in Hebei, Inner Mongolia and Shanxi to supply non-staple foodstuff to Beijing and Tianjin. In the south, the bases would

be set up in Jiangsu, Zhejiang and Anhui to supply Shanghai and other cities in Yangtze River Delta.

9. In the mid-1980s the backwardness of technical equipment was an outstanding problem for small industries, especially township industries. A number of industries still engaged in manual labor with high labor intensity, low productivity and poor quality. As techniques were backward and resources could not be utilized comprehensively, economic benefits were lower and pollution serious. For the development of technical equipment, attention should be paid to its "appropriateness" and its advancement. The equipment should be mature technologically, with a short cycle of development and quick effect. This equipment should also be valuable for extension with notable economic and social benefits. The principle of standardization, seriation and generalization should be carried out in the development of the equipment. It was necessary to try in every way to make the equipment serve more purposes with fewer sets and to avoid repetition and dispersion, to facilitate the arrangement of production.

10. A training plan was undertaken.

The contents of the Spark Program show that all the projects developed were based on local resources and appropriate for local conditions. Insofar as the principle of "short, even and fast" was followed, remarkable effects were brought about in a short period of time. For instance, the project of developing special aquatics (soft-shelled turtle, bullfrog, tortoise) undertaken by the Special Aquatics Research Institute in Hanshou County, Hunan Province, was extended to 1,412 township industries and individuals scattered over 24 provinces and municipalities, including Beijing, Guangdong, Fujiang, Gansu, and Helongjiang.

The Spark Program emphasized comprehensive utilization of the resources of rural areas. Taking projects as a "growth point" and radiating these projects, in a few years the Program brought along the development of a number of important industries in the fields of animal husbandry, aquatics cultivation, valuable fur and leather animals, edible mushrooms and beverages on a larger scale and with good prospects all over the rural areas of the country. For instance, the development of the commercialization of mutton sheep in Inner Mongolia changed the traditional method of animal husbandry, created a whole set of mutton sheep breeds and cultivation technologies, all with less investment, faster effects and shorter cycles. The project provided 66,000 mutton sheep in accumulation and brought along the development of the whole local sheep raising industry. In order to support the long range stable development of these industries, efforts had been made to set up a process of industrial service systems. By so doing, social service organizations were developed with a functioning comprehensive service available under the

Table 13
Achievements of the Spark Program

	Projects Completed (1986–1989 Accumulated)	Output Value Increased (in 100 million yuan) (1986–1989 Accumulated)
State Level	797	60.6
Provincial Level	3,774	81.3
County Level	5,775	79.9
Total	10,346	211.8

Source: "Guide to China's Science and Technology Policy."

orientation of science and technology. For instance, in Ba County of Sichuan Province, with technology backing from Sichuan Agricultural University, a service system process for the development of chicken farms was established. The mechanism of integrating the chicken farm with farmers' households was developed. With the chicken farm taking the lead and the availability of a series of production, supply and marketing services, more than 1,000 households were brought along to develop chicken production. Since the end of the 1980s, 127 Spark Program technology intensive areas and 217 regional mainstay industries have been developed.

The implementation of Spark Program is divided into two phases: the phase of project development and the phase of extension. Only after the project has been completed and, through checking and examination, proved to be a success both technologically and economically can the extension of the project be carried out. The industry or scientific and technical unit, as the developer of a project, is compensated for the dissemination of technology. Actually, extension played an important role in promoting the development of small township industries on a large scale. For instance, Fujian Province had undertaken the development and demonstration of edible mushroom technology as a project of the Spark Program at a national level and succeeded in extending the technology to seven districts, 20 counties and 48 townships and caused the total output of mushrooms to exceed that of Japan, stepping into a new stage of scale economy.

As the Spark Program catered to the needs of local township industries, various localities were enthusiastic in raising capital to carry out the projects of Spark Program. One of the major reasons of the success of the Spark Program (see Table 13) was the initiatives of the localities, especially the township industries themselves. Insofar as they had internal demand for technology progress, they actively took part in car-

Table 14
Composition of Investment of the Spark Program's Projects (Accumulated Investment up to 1989, in 100 Million Yuan)

	Total Amount of Investment	Percentage of Investment
State Investment	1.3	1.0
Local Investment	9.5	7.6
Banking Loan	49.2	39.3
Raised by Local Industries on Their Own	65.2	52.1
Total	115.3	100.0

Source: "Guide to China's Science and Technology Policy."

rying out the projects of the Spark Program and the major part of the investment. It can be found from Table 14 that the investment for the Spark Program projects mainly came from the small township industries themselves as well as bank loans.

The implementation of the Spark Program needed the participation of a large number of scientific and technical personnel. An applicant for a project of the Spark Program should be a township industry with a scientific and technological unit as partner (or as backing) or a scientific and technological unit with a township industry as partner. Of course, it would be best if the applicant could be a research and production combined entity. The Spark Program attracted quite a number of technical personnel. In the 1980s, about 900,000 scientific and technical personnel took part in the Spark Program. However, as it was imperative to bring up township industries' own technical force, the training program was an important part of the Spark Program. In the past decade, 36.80 million trainees participated in various training courses of the Spark Program.

Township industry experienced two upsurges. During the late 1970s and early 1980s, township industry first emerged in the country, with vigor and a momentum of fast growth but with a number of problems. The Spark Program, started in the mid-1980s, played an important role in helping a large number of township industries through the fluctuation by bringing them up to a higher technological level. The second upsurge came after 1992. During that period of time, a large number of township industries, especially those in the southern part of Jiangsu, further upgraded their technological level, improved the quality of their products and scaled up their production; and many of them entered into high-tech fields as well as joined in the export-oriented economy. By the end of 1995, the total number of township industries came up to 6.518 million yuan and the total output value of township industry came up to 3893.35 billion yuan, an increase of 12.5 times over that in 1985. About 200 million people in rural areas had greatly improved their living standards by the growth of township industry. It is estimated that by early next century, the total output value of township industry will account for half the GNP in China.

Currently, township industry is coming under a serious challenge once again, and many township industries are setting up a new goal—to carry out a "second pioneering."

With the development of a market economy, township industry is confronted with challenge from three fronts, as follows.

1. Township industry did have its time on the market at the early stage of the development of a national commodity economy. As the market then was still under the scarce economy, many township industries were able to build good market share by producing products that filled gaps in the marketplace and to enlarge their profit margin by increasing the volume of production and broadening their market share by lowering prices. Their small size and flexibility in changing product lines, as well as low labor and production costs, helped township industry develop at a speed much faster than State-owned industry. Now, with the development of a market economy, such a model of growth is outdated.

2. Compared with State-owned industry, township industry did have advantages in a more flexible operational system in its early stage of development. Township industry was the new child of reform and didn't have the traditional limits of a planned economy. Township industry had been free of administrative control and had operated on its own initiative. However, township industry did not develop according to the conventions of the market economy. For instance, the property rights of township industry were not clarified. It started and developed as a collective industry with the backing of the township. The indistinction in property rights, responsibilities, interests and power couldn't help but weaken the management system and the vigor of township industry as an enterprise.

3. Township industry has been confronted more than ever with structural nonconformity with the development of a market economy. As township industry developed at a low starting point, it inevitably fell into a unitary structure with repetition at low levels, low product mix, low quality of products and inability to develop brand name products, missing the advantage of scale production. Should it be unable to overcome such weaknesses derived from structural nonconformity and fail to realize the upgrade of industrial formation, township industry would be confronted with serious danger.

In view of this, as a "second pioneering," reform of property rights prevails in township industries. Many township industries have transformed into shareholding system or shareholding cooperative enterprises,[2] and some are leased or auctioned and some are merged to transform into enterprise groups. The reform has brought about a better management system and operational mechanism in a large number of township industries. After reform, economy of scale will play an important role in township industry, as quite a number of large enterprise groups will emerge and focus on turning out brand name products. In order to improve the technology contents of products and enhance competitiveness, technology input for the township industry will be increased by a big margin, to accelerate technology renovation and product development.

With the second pioneering, the Spark Program is playing an even more important role in upgrading the technological level and optimizing the industrial formation of township industry. During the ninth five year plan period, the guideline of the Spark Program has been transformed from "short, even and fast" to "high, grouping and outward-oriented." "High" stands for higher technology contents and a higher technological level. "Grouping" stands for the formation of a number of enterprise groups to gain group effects. "Outward-oriented" means that township industry will break up the local market limitation and try in every way to penetrate their products into the markets of other provinces as well as into the international market. Appropriateness was a guideline for yesterday, and "high, grouping and outward-oriented" is the direction for today and tomorrow.

In the coming five years and the period up to 2010, the overall strategy of the Spark Program is:

1. to set up demonstration and direction for the industrialization and scale operation of agriculture;
2. to accelerate the transformation of the economic growth model of township industry into a new model relying on the progress of science and technology; and
3. to strengthen the program of help-the-poor by science and technology.

For the development of township industry, the following five strategic measures will be implemented.

1. The program of marching westward. The development of township industry in different parts of the country is uneven. There is a big gap between the east and the central and western part of China economically.[3] In realizing the goal to improve the well-being of the people all over the country, helping people in the central and western part of China uplift their living standard will be a decisive factor. In the northwest part of China, there are a number of large and medium-sized industries.

These industries were either newly built during the 1950s, when China developed the first batch of large industries with the support of the Soviet Union, or moved there during the 1960s, when the country was developing a defense industry. These industries with modernized equipment were either built in or moved into an underdeveloped area where people mainly engaged in agriculture and animal husbandry. After large industries settled there, quite a few larger or smaller urban cities emerged. However, these cities had no links with the surrounding countryside both economically and socially and resembled isolated islands. Up to now, the cities have remained isolated from the surrounding countryside. However, since resources were there, if the large and medium-sized industries in the cities could provide information, technology and technical equipment and the countryside could contribute their lands, labor force and some capital, the integration of the urban city with the countryside will be realized and a number of township industries will develop. That will be a road to prosperity in the northwest part of the country. With the purpose of developing township industry in the western part of the country and thus bringing prosperity to the people there, the Spark Program is planning to spend 40% to 50% of its funds as guide funds to support the development of Spark Program projects in the western part of the country.

2. Taking existing regional mainstay industries and the Spark Program technology intensive areas as the base, to maximize the important role of industries with regional characteristics and direct the development of modernized small urban townships, with coordinated development of first, second and tertiary industries.

3. Taking high and new technology as the beginning, to focus on the development of regional mainstay industries of the Spark Program, to give prominence to the demonstrative role of economy of scale, industrialization, specialization, grouping effects and internationalization, to propel township industries to carry out scale operations, upgrade product levels and improve economic benefits, to make it possible for the Spark Program to scale new heights.

4. Centering around the construction of regional mainstay industries and the Spark Program technology intensive areas, to carry out the concentrated disposition of management, technologies, capital and personnel, to take advantage of the integration of resources to improve comprehensive benefits.

5. To develop bilateral and multilateral cooperation, to open up channels of cooperation for the internationalization of the Spark Program; to face developing countries and to carry out high level and high quality Spark Program projects, so as to target the international market to develop outward-oriented Spark Program projects.

NOTES

1. In the early stage of the development of township industry, the principle was to encourage the township industries to keep themselves at a small scale and specialize in specific fields with a view to maintaining their flexibility and concentrating their efforts on improving their technological levels. "Combined" meant that these small industries might make up for the limitations of their smallness and specialty by combining.

2. The shareholding cooperative system is usually practiced in smaller enterprises with limited partners. It is a form similar to partnership.

3. In China, there is a big gap between the eastern part and the central and western part in economic development. In 1992, the per capita GNP in the eastern part was 2,800 yuan, while that in the central and western part was 1,517 yuan.

13

Policy on the Development of Infrastructure Industries

As a developing country, China suffered from a lack of infrastructure, which bottlenecked economic development. Therefore in the process of China's modernization, the first priority has been given to the construction of infrastructure. The vast territory and uneven development of different regions of the country has made infrastructure development very complicated. Under the circumstances, the State can only put its financial and material resources into projects of major importance, which makes it even more important to identify focal points for priority development as well as to select technologies.

THE ENERGY INDUSTRY

Energy plays a role as the vanguard of development of the national economy. In the past, when the energy supply failed to meet the demands of economic development, electric power for people's daily use was limited by periodical outages. Some industries were even forced to stop work for three days a week during certain periods of time. Time and again government was reminded to overcome the problem of the insufficient energy supply when the country set off on high-speed economic development. Under the circumstances, priority has been given to developing the energy industry in the past two decades.

In 1997, Li Peng published an important article titled "China's Energy

Policy,"[1] which gave an overall account of the development of China's energy industry and the focus of China's energy policy.

According to the article, by the end of 1996, the primary energy resources produced in China, including crude coal, oil, natural gas, hydraulic power (excluding such energy resources as fagot, methane and wind energy used by farmers' households), converted into standard coal reached 1.26 billion tons, ranking in the forefront of world energy production. However, because of China's large population, per capita energy consumption (converted into standard coal) was only 1.14 tons, still lower than the average level of the world.

China has a large reserve of energy resources, but its geographical distribution is uneven. The coal reserve is abundant, as is the water reserve, but the reserves of oil and natural gas are relatively insufficient. With economic development and the improvement of people's living standard, the demands for petroleum and natural gas will certainly increase by a large margin. Therefore, in addition to endeavoring to develop the reserves of oil and natural gas at home, it is also necessary to utilize foreign resources.

Electric power is an advanced and convenient energy resource. In the past, China frequently suffered from a shortage of electric power. Switching off power to restrict use had been a hindrance to economic development and an inconvenience to the people's daily lives. Since reform, faster progress has been made in the power industry. Now, the increased capacity to generate power is more than 230 million kilowatts, ranking second in the world, and the tension over the power supply has been mostly relaxed. In some areas, power supplies are still insufficient, but in other areas the supply has outstripped the demand.

The Development of Coal Production

China has been one of the few countries in the world to use coal as its main energy resource. According to the statistics, in 1985 the proportion of coal in total consumption of primary energy resources was 30.7% on average in the world, and 23.7% in the United States, 25.3% in the former Soviet Union and 19.7% in Japan; while in China the proportion was 75.9%; in 1996, it was still over 70%.

In China, coal is the power of industry. In 1985, coal-generated energy accounted for 56.9% of electric power, and coal, coke and electricity as well as thermal energy generated by coal accounted for 72.5% of energy consumed by the whole industrial sector.

Coal has been the main raw material of the chemical industry in China. In 1985, coal and coke accounted for 52% of the industry's raw material. China's coal chemical industry has provided agriculture and light industry with a large number of products. In the past decade, this situation has changed because of the growth of the petrochemical industry. Coal has also been the main fuel of people's daily life. In 1985, coal accounted for 91.8% of the commodity energy resources for urban and rural dwellers. In the past decade, the situation has changed with the development of petroleum and natural gas production. In the circumstances, coal production will still play a dominant role in the development of energy in China.

In 1996, China's crude coal output reached 1.38 billion tons, ranking number one in the world. The fast growth of coal production was due to the policy of maximizing the initiatives of both central and local governments and carrying out the principle of simultaneous development of large, medium-sized and small coal mines, which conformed with the national conditions of China. Currently, the ratio of coal output at the central level and local level is 40% to 60%, and local coal output is primary. It is advantageous for small and medium-sized township coal mines to use dispersed odd coal resources and the abundant rural labor force. Peasants run small mines with capital they have raised on their own, with little investment and a pretty good payoff. The development of small mines favors the growth of the rural economy. Currently, the total output of township coal mines has risen to 610 million tons, accounting for about 47% of the total coal output of the country. The weaknesses of township coal mines are a high accident rate, low coal resource recovery rate, higher waste of resources and a weaker sense of environmental protection, which leads to serious pollution. Due to the specific circumstances in China, where there is a large amount of dispersed odd coal resources, small and medium-sized coal mines will eventually develop and play a specific role. As for the weaknesses, they will be overcome mainly by the development of scale operations and technological renovations. During the ninth five year plan period, it is planned that the annual output of a single mine will come up to more than 100,000–150,000 tons and the equipment up to semi-mechanization. On a national scope, more than half the township coal mines have been required to come up to a semi-mechanized level and a certain level of scale operation.

State-owned large and medium-sized coal mines are the main force of the coal industry. The total output of State-owned large and medium-sized coal mines reached 530 million tons in 1996, which accounted for 39% of the total coal output of the country. Though the proportion of the output from State-owned large and medium-sized coal mines had been lowered because of the rise of township coal mines, the former are comparatively advanced in equipment, have higher productivity, higher quality, more varieties and reliable safety measures, which make it possible for these mines to maintain stable production.

Right now, in China, more than 70% of energy is from coal. This situation will probably remain unchanged for a long time. It is estimated that during the ninth five year plan period, the average growth rate of coal will be 2.3%. The total output will come up to 1.45 billion tons by the year 2000, including 50 million tons for export. With technological progress, the decrease of coal consumption, its replacement by other energy resources, and environmental protection, the elasticity coefficient of the growth of coal—the ratio between the growth of coal and the growth of GNP—will come down from 0.38 currently to 0.32. It will be great progress. In the meantime, more coal will be transformed into electric power, gas and even liquefied fuel, such as water liquefied slurry and liquefied oil, and provide the society with cleaner energy resources.

The technological level of State-owned large and medium-sized coal mines has improved greatly in the past decade. Large comprehensive mining equipment has been extensively adopted, which ensures the improvement of productivity and safety of coal production. A single set of large-sized comprehensive mining equipment, operated by more than 100 workers, is able to turn out an annual output between one and two million tons, and sometimes reaching three million tons. Currently, as China can manufacture the entire range of comprehensive mining equipment, more will be used to equip coal mines to promote technology renovation.

In China, 90% of coal had been combusted directly. The obsolescence of technology and equipment made for low thermal efficiency and high pollution. In the past decade, coal consumption has doubled. Many efforts have been made and will continue to be made in the research and exploration of new paths to improve thermal efficiency and reduce pollution. The main contents of research are as follows:

1. New technology of coal washing and formed coal process
2. Recycling fluid-bed boiler power generation technology
3. Coal gasification combined recycling power generation technology
4. New technology of coal gasification
5. New technology of coal liquefaction
6. Large scale coal comprehensive utilization project
7. Economically efficient gas cleaning technology

The Development of the Petroleum Industry

The industrial base of petroleum and natural gas has been relatively weak in China, having been developed from nearly nothing since the founding of the Republic. In 1996, crude oil output reached 158 million tons and natural gas 20.1 billion cubic meters. Even so, the production of oil and natural gas was still unable to meet the demands of economic development. Now, China has become a petroleum import country. It is estimated that during the ninth five year plan period, import will increase.

In order to solve the problem of a shortage of oil and natural gas, it is necessary, first of all, to develop and utilize China's own oil and natural gas resources. Li Peng pointed out in his article "China's Energy Policy" that, according to ascertained oil and natural gas resources and the layout of oil production, a general direction of "consolidating oil production in the Eastern part of the country and carrying out the development in the Western part" should be pursued.

In 1996, in the eastern part of the country, the output of the Daqing and Liaohe oil fields reached more than 120 million tons; in the western part of the country, the output of the Klamayi and Tarim oil fields of Xingjiang reached 20 million tons; and the output of offshore oil was more than 15 million tons.

Daqing oil field has achieved a stable output of crude oil of over 50 million tons per year for the past 21 years. By adopting advanced technology and exploitating a new reserve, Daqing will strive to maintain its current output level for another 15 years. Other oil fields in the eastern part of the country must also rely on technological progress to improve their recovery rates and constantly look for new reserves to maintain their current output.

Rich oil and natural gas resources have been discovered in the western part of the country. There are good prospects in the basins of Tarim, Turpan, Sungar in Xingjiang and Caidamu in Qinghai. The constructing of the Lanzhou-Xingjiang Railway has prepared good conditions for transporting crude oil from the western part of the country. When the output increases to a certain level, an oil and natural gas pipeline will be built to transport oil and natural gas to the eastern and southern part of the country.

The development of natural gas has been at a low level in China. The scale of production is smaller and the ratio between natural gas and oil output is lower than the world average. With efforts made in the past few years, three natural gas fields, including Chongqing, Shanxi-Gansu-Ningxia and Xingjiang, have come into shape with an ascertained natural gas deposit of 1.3 thousand billion cubic meters in accumulation. The basis for accelerated development is available. The natural gas pipeline from north of Shanxi to Beijing has been built. The technology to replace gasoline and diesel with natural gas as fuel for the automobile is matured. Only if the refueling station systems in cities and along the highway can be set up in the natural gas production areas, can the replacement of oil by natural gas be gradually realized.

China's offshore oil and natural gas development was the outcome of reform and the opening up policy. In the mid-1980s, offshore oil development was opened to the outside world, and foreign investment and technology was absorbed. The form of risk surveying was adopted in the cooperative recovery of China's offshore oil and natural gas with foreign firms.

According to the statement of Li Peng on developing the petroleum industry, China is based on domestic resources and goes globewide and will utilize domestic and international resources as well as market resources. China is ready to actively carry out cooperative endeavors with foreign countries in multiple forms, such as taking part in the surveying and exploitation of overseas oil and natural gas resources, contracting overseas projects and providing labor service and establishing a stable import and export market. This is to ensure a long standing and stable oil and natural gas supply for the country.

The Development of the Power Industry

The direction for the development of the power industry is to adapt measures to local circumstances, to carry out development of both hydraulic and thermal power simultaneously, to develop nuclear power appropriately and to carry out synchronic development of power networks. From 1979 to 1996, generated power increased 8.3% per year progressively on average and basically kept growth synchronic with economic development. During the eighth five year plan period, the installed capacity increased by more than 15 million kilowatts per year. The high growth rate of China's power industry is owed to the breakup of the original practice of running the power industry under a single administrative department. Instead, a new system was set up that let the power industry run with multiple departments, and capital was raised through diversified channels. In so doing, aside from the central functionary department, the initiatives of local governments and enterprises and both the domestic financial institutions and international investors were brought into play. However, due to the characteristics of the power industry, power supplied by individual power plants cannot ensure a stable supply for the end users. It is necessary to carry out a power supply through unified power networks. Therefore, while encouraging multiple departments to run the power industry, it is necessary for the State to run and manage the power networks. Currently, six transregional networks and a number of networks at the provincial level have been formed. Transmission line connections are set up between transregional and provincial level networks to improve the safety and reliability of the power supply and gain economic benefits through networking. In the future, the construction of the Three Gorges Hydraulic Power Station will be an opportunity to promote the gradual formation of a nationwide unified network. The establishment of big power networks makes it possible to adopt large volume and high parameter generator sets and save on unit investment, as well as to gain higher efficiency and a shorter construction cycle. However, the construction of big power networks also brings about management difficulties and requires advanced technology to manage the networks to ensure safety. Sophisticated information technology will certainly be necessary in extensively applying network management, to ensure the safety of networks and improve economic benefits.

Less investment and shorter cycles are needed for the construction of thermal power. During the ninth five year plan period, thermal power will account for about 75% of the total power. However, thermal power causes environmental pollution and requires a high volume of coal transport. In the future, it will not be adequate to continue to develop more thermal power in large and medium-sized cities. China should construct more electric power stations near coal mines and, at the same time, continue R&D on cleaner coal power generation technology.

China has abundant hydraulic power resources, and hydraulic power is a clean and renewable energy resource. However, the construction cycle of hydraulic power is longer and investment greater, and it is often comes with the complication of removing a large number of local dwellers. Therefore, hydraulic power has remained at a low proportion of energy resources. Of course, the situation should be changed. The Three Gorges Hydraulic Power Station will be the largest one in the world when it is completed. Its installed capacity will be 18.20 million kilowatts. The largest water conservation pivot project on the Yellow River, the Xiaolangdi Dam,[2] is also under construction. The hydraulic power station to be built will also be large.

During the ninth five year plan and the first decade of the next century, the power industry will maintain a certain growth rate. According to calculations, the newly increased installed capacity in the ninth five year plan period will be 80 million kilowatts, or 16 million kilowatts per year on average.

Large scale power construction requires substantial investment, and capital will be raised from multiple channels. Insofar as there is a stable market for the power industry, it will be able to look to both domestic and international financial markets for financing. Overseas capital may be absorbed in diversified and flexible forms for constructing and running power plants, such as equitable joint ventures or contractual joint ventures or by way of BOT (build, operate, transfer). The power industry is a basic industry of the State with a comparatively stable investment return rate. For large-sized China-foreign joint venture power plants, the Chinese Party should keep the necessary share of control.

Nuclear power is a safe, reliable and clean energy source. The record of nuclear power operations in the world has proved that the few different types of nuclear power reactors, such as pressurized water reactor, boiling water reactor and heavy water reactor, are all safe and reliable. China is a nuclear country, but has had a late start in nuclear power development. The pressurized water reactor, which has been used frequently among the various reactor types in the world, has been selected as the main reactor type to be used in China. Currently, three generator sets are being operated in Dayawan and Qinshan, with a total installed capacity of 2.10 million kilowatts. These account for about 1% of the total generated power in the country. During the ninth five year plan period, due to limited capital, nuclear power will not experience huge development. There are four nuclear power projects, including eight generator sets with a total installed capacity of 6.60 million kilowatts, under construction. These projects will be completed by the tenth five year plan. By that time, the proportion of nuclear power to total generated power will be about 2%, still only a supplemental energy source in China.

China has rich nuclear energy resources, certain technological capabilities and a qualified research and technical force for the development of the nuclear power industry. There are good prospects for the development of nuclear energy. In order to make stable development possible for the nuclear power industry, it is necessary to take the following three steps.

1. Utilize foreign capital to purchase foreign equipment and technology at the start of China's nuclear power industry. The success of Dayawan Nuclear Power Station is a good example. Seventy percent of the power generated by Dayawan Nuclear Power Station is transmitted to Hong Kong. The project has better economic benefit, and at the same time, China has learned the technology of constructing a large scale nuclear power station.

2. The principle of rolling development may be applied in the development of the nuclear power station. Its construction requires substantial investment and a longer construction cycle, but the cost to generate power is not high. During the later stage of repaying principal and interest, it will be able to produce a comparatively high profit. After repayment of principal and interest, more profit will obtain. Profits can then be used for the rolling development of nuclear power.

3. To carry out scale operations.

TRANSPORTATION

Just as it is important for blood to circulate in human body, in a country as large as China the circulation of people and freight is of major importance to social and economic life. In addition to passenger and freight transportation in general, the transportation of energy resources has been a heavy burden for China's transportation system. Coal, being the dominant energy resource, must be transported from the western part of the country to the eastern part in high volume. Most of the coal reserves, as well as large State-owned coal mines, are located in the western part of the country, and coal transportation is chiefly through railway. Therefore, the railway system has long been overburdened with the major part of transportation of passengers, freight and energy resources. The guideline for developing transportation is to improve transportation capacity and gradually readjust the composition of the transportation system, to maximize the superiority of various transportation means and to establish an economically reasonable transportation system with coordinated development of various transportation means.

Therefore, aside from developing a railway system, which is still important for the expansion of overall transportation capacity, it is imperative to divert the workload of the railway system to other means of transportation. It is important to accelerate the development of a highway system, civil aviation system and water transportation system. The development of a highway system will divert near distance freight transportation, the development of a civil aviation system will divert long distance passenger transportation and the development of ocean shipping will divert long distance staple material transportation.

Since the readjustment of the transportation composition, the situation has improved, but the overburdening of the railway hasn't been thoroughly overcome yet. The potential for water transportation, especially that of inland navigation, still need to be tapped. In the meantime, oil pipeline transmission has to be developed further.

The capacity of energy resource transportation must be expanded and new technologies developed for energy resource transportation.

In China, energy resource transportation has been an important part of railway transportation. It is one of the focal points of the technology policy of the transportation industry that multiple forms of heavy-load transportation should be developed to help solve the problem. In the light of the policy, high-power locomotives and large freight trains have been used. The Datong-Qinhuangdao[3] electric railway line was exclusively designed as a special line for heavy load coal trains. The ocean and continental combined transportation has been a big step in the development of coal transportation. The coal from Shanxi Province carried to Qinhuangdao by train was transferred to the south by ship.

Rapid growth has taken place in the development of pipeline transmission of crude oil and natural gas. In East China, a crude oil pipeline network, which connects six oilfields, directly supplies 14 refineries and leads to four ports including Dalian, Qinhuangdao, Huangdao and Yizheng, has been formed and used for closed transmission. Progress has been made in improving the transmission technology of high solidifying and high viscosity crude oil and reducing the consumption of energy.

Table 15
Development of the Railway System in China

Year	Weight/Length Unit: 100 million tons/km	Person/Length Unit: 10,000 persons/km
1992	11,620	3,150
1993	11,936	3,582
1994	12,462	3,637
1995	12,900	3,547
1996	12,895	3,357

Source: The Statistical Bulletin on the National Economic and Social Development of the PRC by the State Statistical Bureau.

Now, it has been realized that pipeline transmission and combined pipeline and water transportation become the main way for crude oil transportation. The pipeline transmission of coal is under active and careful development, and the construction of a pipeline transmission of finished oil is being planned. Nonetheless, the growth of transportation capacity still lags behind the growth of the energy industry, especially that of coal production.

Railway transportation must be further developed. The railway is playing and will play a dominant role in China's transportation industry (Table 15). Huge efforts have been made to build new railways on the trunk lines of communication. Jing-Jiu (Beijing to Kowloon) Railway finished construction in 1995. It has a total length of 2,536 kilometers, running across nine provinces and large cities, and is the longest railway with the largest investment and had the largest construction scale in the history of railway construction in China.

The acceleration of the renovation of the traction force is an additional way to develop railway transportation. Progress has been made in developing electric as well as internal combustion traction railways. As early as 1987, the total force (in tons/km) of electric traction and internal combustion traction together reached 55.3% of the total traction force. It marked a new epoch in China's railway industry by taking electric and internal combustion traction as the primary focus. In the process of renovating traction force, the internal combustion traction was developing faster than that of electric traction. Up to 1987, the total tons/km of internal combustion traction was 3.7 times that of electric traction. This deviated from the principle of "taking electric traction as the main in railway traction." Therefore, a general orientation is to further accelerate the development of electrified railway.

A big effort was needed to develop containerization and grain and

Table 16
Development of the Highway System in China, 1992–1996

Year	Weight/Length Unit: 100 million tons/km	Person/Length Unit: 10,000 persons/km
1992	3,500	3,030
1993	4,175	3,582
1994	4,481	4,147
1995	4,890	4,726
1996	5,000	5,060

Source: The Statistical Bulletin on the National Economic and Social Development of the PRC by the State Statistical Bureau.

cement bulk transportation as well as cold-storage transportation. There has been big development in containerization in China. Internationally, China has developed ocean international containerized transportation to most overseas countries. The railway and highway transport departments have carried out dispersal transportation of international containers from the ports. There is also a big development of domestic containerized transportation. Container technological equipment has been much improved, and a large number of special container ships and a number of special container berths have been built.

The development of grain and cement bulk transportation and cold-storage transportation has been slower than that of containerization. In particular there were quite a few difficulties in the development of cold-storage transportation. Aside from the requirements of the railway refrigerator wagon, refrigerator truck and refrigerator ship, cold-storage transportation also needs to pre-cool in the places that grow fresh fruits, vegetables, aquatic foods and perishable goods. The weakness of the pre-cooling process leads to an inability to form a "cold-storage chain," which makes development of cold-storage transportation slow. The situation has improved in recent years, but it will still take a great deal of effort to accelerate its development.

The improvement and construction of highways and the development of motor transport must be accelerated. The construction of a new highway system saw rapid progress in the past decade (Table 16). By the end of 1987, the total highway length throughout the country was 982,000 kilometers. Since then, newly constructed highway increased by about 2,000 kilometers per year on average. By the end of the century, the total length will come to 1.25 million kilometers, including more than 8,000 kilometers of superhighway. In 1996, there was further progress in highway construction. Newly constructed highway came up to 25,000 kilometers in that year, including a 1,117 kilometers superhighway. In the

Table 17
Development of Water Transportation in China

Year	Weight/Length Unit: 100 million tons/km	Person/Length Unit: 10,000 persons/km
1992	13,300	180
1993	13,672	205
1994	15,704	175
1995	17,000	180
1996	17,760	168

Source: The Statistical Bulletin on the National Economic and Social Development of the PRC by the State Statistical Bureau.

past few years, with a view to making up for the shortage of funds for the construction of communication facilities, aside from raising capital domestically, China continued to use loans from international financial institutions and foreign governments, as well as carrying out BOT (build, operate, transfer) projects as an experiment. By the end of 1996, China had signed 27 projects with the World Bank for a total loan amount of U.S. $4.2 billion. Starting from the eighth five year plan, China has planned to spend 30 years to complete the construction of a main skeleton structure of the nationwide highway networks.

It is necessary to develop sea and inland river navigation (Table 17). For sea navigation, both the construction of ports and development of a sea transport fleet are important.

Since various kinds of ports have been constructed, the throughput capacity of ports has increased by a big margin. Special ports loading and unloading capacity for coal, bulk grain and timber have increased by a large margin. The throughput capacity of ports in Shanghai was over a hundred million tons as early as the end of the 1980s. This made Shanghai one of the 10 largest ports in the world. The throughput capacity of main coastline ports in China reached 835 million tons by the end of 1996.

The ocean transport fleet has seen rapid development. As early as 1987, the carrying capacity of China's ocean fleet reached 2 million tons, ranking eighth in the world. The technology composition of the fleet has been improved and transformed into a multipurpose and modernized fleet. There has been an increase of container vessels and roll-on, roll-off ships. The proportion of domestically made vessels has increased.

By way of technology importation and export of ships made in China, China's ship-building industry has come up to international standards. Breakthrough have been made in the technology of designing, construction, and manufacturing of equipment as a whole set, which has made

Table 18
Passenger Turnover of Civil Aviation in China, 1992–1996

Year	Unit: 100 million passengers/km	Growth over Previous year (%)
1992	399	32.5
1993	515	26.9
1994	533	11.6
1995	652	18.1
1996	752	10.4

Source: The State Statistical Bulletin on the National Economic and Social Development of the PRC by the State Statistical Bureau.

the vessels made in China a commodity with international competitiveness.

The speed of development of inland river transportation has been slower than that of ocean transportation. Waterway construction is the base for the development of inland river transportation. The communication and transportation department had mapped out a channel construction plan and waterway classification standard. However, inland river waterways and ports in the country have been basically in a natural state. As inland rivers haven't undergone comprehensive development for a long time, the potential of inland river transportation has remained untapped. When the Three Gorges Project is completed, vessels of 10,000 tons will be able to get through Yangtze River and open a new chapter in the inland river navigation history of China.

The development of air transportation is of special importance in absorbing long distance passenger transportation. Less investment and a shorter cycle are needed to develop an air traffic system than to build a railway line. The payoff from air traffic is faster, and the occupancy of land, compared with building a railway line, is insignificant.

In the past decade, civil aviation saw a fast development in China (Table 18). In order to develop civil aviation quickly, it is imperative to satisfy the demand for mainline aircraft. China's aeronautic industry started in 1951 and formed a comparatively complete industrial system in the 1960s. However, for a long time, it had engaged mainly in the production of military aircraft. In recent years, with the readjustment of the direction of development of the defense industry, the aeronautic industry has been transformed to "combining military use with civilian use and taking civilian use as the main" to adapt to the requirements of constructing modernization. In respect to the composition of demand, orders for military aircraft have been reduced drastically, while demand for aircraft used for both freight and passenger transportation and for general purpose has increased. Therefore, the key to rejuvenating China's aeronautic industry is the development of aircraft for civilian use, especially the development of aircraft used for flights on main airlines. The domestic-made "Yun No. 7" aircraft has been used in a number of feeder lines, but those used for flights on the main airlines are still either bought or leased from foreign countries.

In order to meet domestic demand, China should accelerate research and development on the aircraft used for main airline flight. According to the current R&D capability in China, it is necessary to introduce from abroad mature technology for aircraft for main airline flights. Entering into cooperative production with an international manufacturer of approved aircraft should be considered for main airline flight, using domestic-made parts and components in the assembly of aircraft, and manufacturing should be localized gradually. China's cooperation with McDonnell-Douglas is an example, although just a start. China may also use her market to exchange technology. Nonetheless, China should focus on doing advance research on key technological issues and, in accordance with the principle of "advance research plus cooperation," gradually make breakthroughs on key issues to improve the level of R&D.

As airworthiness regulations are an important guarantee of the safety and comfort of modern civilian aircraft as well as a decree to be observed, China is persistent in the implementation of international standards. Both the MD-82 aircraft assembled by Shanghai Aircraft Corporation and parts and components produced by Xian Aircraft Corporation for Boeing aircraft have obtained certificates issued by the FAA and therefore can enter the international market. The implementation of international standards in China's aviation industry means the necessary renovation of existing production conditions, technological conditions and management systems in enterprises. It also means conformity with a whole set of international standards, from raw material standards, designing standards, industrial standards and experiment standards to maintenance standards and management systems. It takes much effort and substantial investment. It had taken seven years and an investment of about 10 million yuan to renovate the production line to obtain FAA certification for a few small and medium-sized parts and components supplied for Boeing aircraft. To achieve international standards, the related industries of aircraft manufacturing should also implement international standards. The review and examination according to international standards is not limited to the aviation industry itself, but also covers externally supplied raw materials, basic elements and accessories. Therefore, R&D and the production of modern aircraft used for main airline flight represent the industrial level of a country. It has a long way to go.

Table 19
Growth of Telephone Spread Rate in China, 1978–1996

Year	Spreading Rate	Increase of Percentage
1978	0.38	
1987	0.75	Nearly double the spreading rate in 1978
1994	3.20	
1995	4.66	Increase of 1.46 over the previous year
1996	6.33	Increase of 1.67 over the previous year

Source: The Statistical Bulletin on the National Economic and Social Development of the PRC by the State Statistical Bureau.

TELECOMMUNICATIONS

The principle of the technology policy for the telecommunications industry is a high starting point for the adoption of new technology. Under the principle, China's telecommunications industry has seen rapid development in the past decade. It skipped over technological stages on which developed countries spent decades and directly leaped forward to modernized technology. The technology development strategy helped China to spend only about a decade, starting from a low technological level, to catch up with the rapid changing telecommunications technology in the world.

The growth of the telecommunications industry in China is extraordinary. In 1987, the total telephone switching capacity was about 8 million lines, and by the end of August, 1997, the total telephone switching capacity reached 100 million lines. Now China has the second largest telephone network in the world, not very far behind the United States.

Of course, due to the large population, the spread rate of telephones in China does not compare to that of the United States. However, it has seen tremendous growth in the past decade (Table 19).

In recent years, the mobile telephone also saw rapid development in China. During the period 1992–1996, the country's total number of mobile telephone subscribers expanded at an annual average rate of 17%. China has now the world's third largest mobile telephone network, after the United States and Japan.

Guangdong Province is one of the major powerhouse, leading the nation in the modernization of the telecommunications system. The total number of telephone subscribers in Guangdong (including mobile telephone users) reached 10 million by the end of August, 1997, the largest number among all provinces. Guangdong's mobile telephone users number 2,774 million, the largest in the country. The province is working to bring the total number of phone subscribers to 15 million by the year 2000. The average phone coverage rate province-wide reached 16.4% compared with 7.3% nationwide. The coverage of the urban areas of the province has reached an average of 46% and is over 60% in the cities of Guangzhou, Shengzhen, Zhuhai and Shandou. China's average urban telephone coverage was 25% by the end of August, 1997.

Another powerhouse, Shanghai, had a total telephone switching capacity of 5 million lines by early September, 1997. During the past five years, Shanghai has invested 3 billion yuan to build and upgrade its mobile communications system.

The experience of the rapid growth of China's telecommunications industry has proved the important perspective of the technology policy. In the 1980s, China experimented in three districts in the Zhejiang Province with the effects of the different composition of technologies. The result was that the district completely adopted digital technology, digital transmission and digital programmed exchange equipment was not only higher in technological level and better in communication quality, but also had notable economic benefits. The experiment encouraged the telecommunications department to adopt the new technology during the start of the development of the telecommunications industry. It turned out to be the right decision, which helped the country to use a good opportunity to catch up with the world advanced level, without accommodating the original technological base.

Though China has had rapid growth in the telecommunications industry, taking the large population into consideration, the nation still has a long way to go in developing telecommunications. Nonetheless, the current technological level provides a good base for further development. By the end of 1995, in respect to long distance telephone circuits, the digital circuits accounted for 92%. By the same time, 280,000 km optical cable had covered the whole nation. The programmed exchangers accounted for 99.74% in total local automatic exchangers and 99.93% in total long distance automatic exchangers.

Since reform and opening up, China has developed a series of joint ventures to manufacture programmed exchangers, optical cable, digital terminals, mobile telephones, transmission system and so on. The domestic R&D capability has much improved, and a series of high-tech products have been developed, such as SP 30 and 04 programmed exchangers, 622 Mb/s SDH series and 2.4 Gb/s SDH series equipment for optical communication, wavelength modular multiplex experiment system, 155 Mb/s SDH digital microwave experiment system and GSM system.

In order to promote the accelerated development of the telecommunications industry, aside from technological development, it is imperative to improve the operation system. In China, telecommunications had long been under a State monopoly. Only the Ministry of Post and Telecommunications (MPT) had the right to run the telecommunications business. A few years ago, with a view to promoting the development of a telecommunications system and to improving service, the United Communications Corporation (UNICOM) was approved to enter the telecommunications market. It helped to introduce the mechanism of competition into the telecommunications market; as a result, the price of the mobile telephone has been brought down by a big margin. However, competition in the telecommunications market is limited. Up to now, UNICOM's mobile telephone system has only 200,000 subscribers, while MPT's mobile telephone system has reached 10 million subscribers. As a government department, MPT has monopolized communications resources. UNICOM had completed the construction of the municipal telephone network in a few large cities, including Beijing, Tianjin and Chongqing, but was unable to put these into operation for a period of time, as a series of key issues such as interconnection with the public telephone network, communications resources and telephone charges still haven't been resolved in a reasonable way. In view of this, the news media are appealing for the liberalization of the telecommunications market. It is analyzed that there are three important positive factors for faster development of the telecommunications industry. First, there is a greater social demand for fast development of the telecommunications industry; second, according to the macro system of the State, investment for the development of the telecommunications industry from various sectors is welcomed and third, capital from various sectors is actively trying to get into the telecommunications market. However, all these activities are frustrated by the monopoly of the telecommunications department.

The traditional telecommunications department has been both playing the role of a government functionary department and engaging in market operation. Under such circumstances, it is easy for the department to keep its monopoly status by issuing various "policies" and "regulations and rules" in favor of its own operations. For instance, according to the opinion of the news media, it doesn't sound reasonable for the telecommunications department to charge mobile telephone users a very expensive "network entry fee" and phone subscribers a "first installment fee." Since the national public network has been established with the money of taxpayers, it should be fully utilized by taxpayers and shouldn't get a return on investment from the consumers.

The news media proposed to promote competition in the telecommunications market by introducing more modern telecommunications firms into the market, including firms started by private capital.

The appeal of the news media has a general background. In Jiang Zeming's report at the Fifteenth National Congress of the CPC, it was pointed out that industrial monopolies should be gradually phased out. Therefore, breaking up a monopoly and introducing the mechanism of competition into the telecommunications sector will be a general trend that will certainly help promote the further development of the telecommunications industry. The latest development is that China's State Council has carried out an important institutional reform that has led to the cancellation of a number of ministries, including MPT, the Ministry of Electronic Industry and the Ministry of Broadcasting, Movie and Television Industry; and, as a replacement of the three ministries, a new ministry, the Ministry of Information Industry, was established. The latest development of institutional reform will surely produce a positive influence in breaking up the telecommunications monopoly and spur an even faster growth of the telecommunications industry.

NOTES

1. Li Peng, former Premier of the State Council of the PRC. Li's article "China's Energy Policy" was published in *Qiushi* (Seek Truth), No. 11, 1997.
2. Xiaolangdi is located in Henan Province. The Xiaolangdi water conservancy pivot engineering project is a major project carried out on the critical spot of the main stream of Yellow River passing through Henan Province.
3. Datong, one of the most important cities, with abundant coal reserves and large coal mines, is located in Shanxi Province. Qinhuangdao, one of the most important ports in North China, is located in Hebei Province.

14

"Mainstay Industries" and Technology Policy

Since the early 1990s, with the further development of the opening up policy, more foreign businesses have entered China, which has brought China both opportunity and challenge. Confronted with stronger competition, China has had to develop her own intellectual property rights and mainstay industry.

The machinery and electronic industry, petrochemical industry, automobile industry and construction industry have been identified as the four mainstay industries in China.

THE MACHINERY AND ELECTRONIC INDUSTRY

The level of production technology and economic benefits of various sectors of the national economy is determined by the performance, quality, efficiency and speed of renovation of machinery and electronic equipment. Therefore, the machinery and electronic industry is of primary importance in the mainstay industries.

As early as 1986, when the focal points of the technology policy in 12 industrial sectors were issued by the State Council, guidelines for technological development of machinery industry were publicized. The technological development in the machinery industry focused on the improvement of basic parts and components, basic technologies and basic machinery as well as the enhancement of the capability of manufacturing equipment on the whole.

The weakness in production of basic parts and components in China has a historical background. Historically, China's most basic parts and components industries were subordinated to principal machine industries and separated from the latter at a later stage. Since the general guideline for developing the machinery industry in China had long been "to lay emphasis on manufacturing equipment on the whole and bring along the development of basic parts and components industry," there hadn't been an overall plan for the development of basic parts and components. The basic parts and components industries were scattered all over the country and each department and region had developed its own basic parts and components industries and tried to exercise its own regional protectionism. Certainly, the situation didn't help promote the sound development of the industry, but only led to the repetition of constructing basic parts and components industries at a low level. Many basic parts and components industries were working with obsolete equipment, under a low management level and with a weak technical force without the capability to organize volume production of high performance basic parts and components. Therefore, over a comparatively long period of time, the development of basic parts and components industry had been listed as a strategic focal point for the development of the machinery industry. In the past decade, since special attention has been given to developing the basic parts and components industry, the situation has much improved. By the late 1980s, through tackling key technological issues, technology development as well as technology acquisition, assimilation and absorption, thousands of new basic parts and components had been developed in five or six years and most part of such new products had reached the level of the late 1970s and early 1980s internationally, which brought approximately 35% of the main basic parts and components up to a new level. For instance, oil leakage of hydraulic pressure parts and components had been a serious problem in the past, and now the problem had been solved for 70% of the hydraulic pressure parts and components. However, the improvement hadn't brought the basic parts and components industry to a comparatively advanced level in terms of quality and performance. Therefore, it is still a focal point for the development of machinery industry to further improve the basic parts and components in the 1990s.

The development of basic technologies in the machinery industry had long been lagging behind, as industrial enterprises paid more attention to near-term benefits and less to the improvement of basic technologies that might improve the quality of products fundamentally but was unlikely to bring about any immediate benefits. In the past, under the planned economy, the industrial enterprises were under various central ministries and regional administrations, while the institutions devoted to the research of basic technology were separated from industrial enterprises and directly subordinated to various administrative departments. Insofar as the production tasks were assigned by the senior authority just on a product basis, the research institutions were unable to get any assignments and, hence, had no funding. The industrial enterprises, when receiving the assignment of production tasks, didn't like to use part of their production funds for research of basic technologies. This couldn't help but lead to narrowing the scope of basic technology research. Since the mid-1980s, the development of basic technology has been one of the focal points of the technology policy of the machinery industry.

Upgrading the technological level of basic machinery has also been a focal point of the development of the machinery industry. The focus of developing basic machinery is to upgrade the manufacturing technology. In China, it is of special importance to follow the development of manufacturing technology of a new generation, and the core is the development of automation in the manufacturing industry.

The path of automation taken in China has been similar to that of the developed countries—from a rigid system to a flexible system. Early in the 1950s, China introduced a number of rigid automation systems from the Soviet Union and applied them to automobile and tractor manufacturing. Of course, there had been a big gap between China and developed countries in respect to the extent of automation.

As early as the seventh five year plan (1986–1990) period, the introduction and assimilation of a flexible manufacturing system, industrial robot and numerical control machine tools as well as R&D on laser processing machine tools had been listed as major technological issues to be tackled. Now, the basic systems representing the manufacturing technology of a new generation have gradually been extended to China's manufacturing industry. For instance, in recent years, CAD technology has had rapid development in China and is now used extensively in industry. In addition, in order to adapt CAD technology to the applications of Chinese engineers, through secondary development, with the Chinese independently developed CAD system, Chinese engineers are able to access the CAD system with a week's short training and thus help promote the extension of CAD technology. Now, CAD software developed on China's own accord with China's own intellectual property rights has appeared on the market.

The computer integrated manufacturing system (CIMS), as a new model for industrial organization, management and operation, has been included in China's 863 High Technology Research Program. The result of the 863 Program has been industrialized and applied to about 70 industries, including machine-building, electronics, telecommunications, aeronautics, aerospace, light industry, textiles, petroleum, chemicals, metals, coal mining, ship-building, and weapons. Beijing First Machine Tool Factory was one of the pacesetters in developing CIMS. The Factory didn't pursue the automation of the bottom line equipment, but focused on information integration and the improvement of the operational mechanism. In order to realize integration of the control of information flow and material flow among various sub-systems, the Factory adopted two measures: one was to realize automatic paperless processing for box parts and components from designing, technique formulation and NC programming through production scheduling; the other was to realize low-cost CIMS integration in most workshops (i.e., the bottom line equipment of these workshops was constituted of ordinary machine tools and progress was reported artificially to MRP I). By adopting the CIMS system, the Factory had succeeded in shortening the delivery of super-heavy milling planer within 18 months (the original delivery time was 36 months) and thus winning international bids. The efforts in developing CIMS will be furthered to bring China's machinery industry up to a new level.

Automatic monitoring and control is an important field in automation and also one of the focuses of the development of the new manufacturing technology. There is also a big gap between China and developed countries in this field. In recent years, China has tried hard to catch up to the rapid development of automatic monitoring and control technology. For instance, in the iron and steel industry, the production campaign of the blast furnace is directly decided by the quality of the furnace lining. Previously, since it was impossible to make sure of the speed of corrosion of the furnace lining through direct observation, necessary measures had not been taken to regulate damage to the furnace lining. Now, a new on-line monitoring device on the thickness of furnace lining has been developed, which helps extend the production usefulness of the blast furnace. The progress of the adoption of new manufacturing technology needs to be made step by step, as it relates to applications in various industrial fields.

The improvement of the capability in manufacturing major equipment on the whole has been a focal point of technological development in the machinery industry. In the 1980s, China made progress in this regard. Through R&D, China's own machinery industry was able to undertake the manufacturing of quite a few major sets of equipment, such as the whole set of large scale open cut mining equipment at the scale of ten million tons, the thermal power plant unit at the capacity of 300,000 kw, the whole set of equipment for 500,000 volt AC transmission and transformation line and the 2030 cold tandem mill and 2050 hot tandem mill used by the Baoshan Iron and Steel Corporation. China was able to provide the whole set of equipment for synthetic ammonia at the capacity of 300,000 tons and urea at the capacity of 520,000 tons. In the 1980s, China still suffered repetition of the importation of equipment in whole sets, especially those that China was capable to manufacture on her own. The notable example was the import of large scale chemical fertilizer equipment. China had imported 17 sets of large scale chemical fertilizer equipment in the 1980s, and more of them were imported in the early 1990s. It was partly due to the shortage of funds domestically and partly to the fact that China didn't have a relatively stable policy on major technology equipment that conformed to China's own national conditions and was accepted by both the departments as users and the departments as manufacturers during that time. In addition, there were still some technical reasons, such as a lack of integration between cooperative design, cooperative production, technology acquisition and R&D on China's own accord. Of course, China needs to further improve her capability in manufacturing technological equipment in whole sets. This is still relatively weak.

As a development goal, it is indicated in the Ninth Five Year Plan for National Economic and Social Development and the Outline of the Long Range Goal up to 2010 of the PRC that in the machinery industry, the emphasis is to upgrade the level of the development and manufacturing of whole sets of large scale power, chemical fertilizer and ethylene production equipment; to improve the quality and performance of major basic machine tools such as numerical control machine tool and the major basic parts and components such as hydraulic pressure, pneumatic drive and sealing parts and components as well as meters and instruments, and, to strive for the development of new products and enhance the capability of manufacturing whole sets of equipment. By 2000, the proportion of numerical control machine tools shall account for 18% of the total output of machine tools, more than 80% of the processing equipment required by manufacturing industry shall be made domestically, and more than 50% of processing equipment required by the sedan car industry shall be made domestically. In respect to the whole set of large scale thermal power equipment, emphasis shall be placed on the development of supercritical and co-cycling units. The production capability of large scale hydraulic power equipment shall be upgraded to five million kw with the capability of manufacturing a large scale power plant unit with a capacity of 700,000 kilowatts per single unit, the pumped-storage set at a capacity of 200,000 kilowatts and the tubular turbine set at a capacity of 40,000 kilowatts.

In respect to the development of electronic industry, it is indicated in the Ninth Five Year Plan for National Economic and Social Development and the Outline of Long Range Goal up to 2010 of the PRC:

Emphasis shall be laid on the development of IC, new devices, new computer and telecommunications equipment, so as to strengthen the capability of providing economic and social development with up-to-date information system and facilities. Preferential policies shall be carried out for the development of IC and special IC projects shall be put into implementation continuously.

To develop micro-chip devices, new display devices, [and] photoelectric devices, and to establish production and export base of computer and accessory devices such as peripheral unit and PC motherboard, etc.

To develop basic software and applied software conformed to international standard. To stress the development and production of the equipment of digital programmed exchange, mobile communication and optical communication, etc.

In 1996, the first year of the ninth five year plan, according to the data provided by the Ministry of Electronic industry, the electronics industry in China had realized a total output value of 298.2 billion yuan, a 200.7% increase over the previous year and a total sales of 194.7 billion yuan, a 18% increase over the previous year.

During the eighth five year plan period, in tackling the key technological issues in electronic industry, microelectronic technology and R&D on IC was the focus. The designing and processing technology of 0.8–1.2 micron IC had come up to the stage of small volume pilot production, and a VLSI designing and manufacturing technology development base had been built in Beijing. Further, 1.5 micron CMOS technological optimization had been realized and transformed to volume production. A number of technology results on IC, CAD, VAT, differential analysis technology and key equipment made it possible for China's microelectronic technology to come to a new stage and became one of the major technology sources of China's IC industry. According to the Outline, during the ninth five year plan, six inch and 0.8 micron IC will be formed into volume industrial production, and, at the same time, the special project of eight inch and 0.5 micron IC will be come into shape and be put into industrial production. R&D on 0.3 micron IC will be started during this period of time.

In the development of the computer industry, China has covered all stages of development, formed an overall industrial setup and laid an industrial foundation with certain industrial scales and strengths. However, the repetition in constructing production lines, the dispersion of investment and limited production scale had restricted the development of the computer industry and kept the whole industry at a low level. It was not until the early 1990s that faster growth had appeared in the computer industry. Though there hadn't been an all-round technological breakthrough in the computer industry, there was a few and partial successes in quite a few fields, such as Chinese character laser photocomposition system, PC motherboards, CD-ROM drives, UPS, and so on. All these successes are related to China's own technological superiority and lower labor cost. Hence, people in the computer industry found that if China could take advantage of her own technological superiority in certain fields and make good use of the opportunity provided by the readjustment of the computer industry internationally, China's computer industry would prosper.

During the eighth five year plan period, a rapid development had emerged in China's computer industry. With the high speed economic growth during the eighth five year plan, the extensive application of the computer system in industrial, financial and commercial sectors as well as in management systems, office work, educational systems, family and recreation, had brought about an ever broader computer market. Many world leading computer corporations got into China's computer market, and some of these corporations had deeply penetrated the market. In the meantime, China's domestic computer industry had rapid growth. In 1996, through reorganization, a number of electronic corporations or groups had further strengthened. In 1996, China's PC market had a volume of 1.80 million sets, an increase of 56% than the previous year. The total sales of computers came to 88 billion yuan, a 30% increase over the previous year, while the market share of the four domestic PC giants, Great Wall, Legend, Tongchuang and Langchao (Tide), had come to 40% altogether. Right now, customers in China can get various mainstream PC products on the domestic market. The effect of scale production is one of the major reasons for the success of domestic PC manufacturers. For instance, China Great Wall Computer Co. Group built its computer manufacturing base in Shenzhen in 1995. It was the first PC development and manufacturing base that had scale production capacity in China. It was the largest computer industrial manufacturing base in China with a total construction area of 50,468 square meters. Of course, in order to create a brand name PC, it is necessary to master advanced technology and management methods. Since 1994, Great Wall Group has developed three joint venture enterprises with IBM to engage in the manufacturing of a high level PC, PC motherboard and magnetic head and magnet resistor. The joint venture enterprise produced both Golden Great Wall and IBM PCs simultaneously. Great Wall Group has developed into a State-owned large scale high-tech comprehensive enterprise group. Its annual sales of PCs has increased from 20,000 sets in 1992 to 200,000 sets in 1996, a tenfold growth in five years. Its annual sales are expected to rise to 16 billion yuan by 2000.

Though the international competition on the world computer market has intensified, people in China's computer industry believe it a good opportunity for China's PC industry to carry on a simultaneous development with the international computer industry. The industrialists involved in computer manufacturing circles believe that this can be realized because of the following reasons:

1. The manufacturing of the PC worldwide is transferring to developing countries. A sound environment for the manufacturing of the overall unit and parts and components of PCs has come into shape in China. In Pearl River Delta and Suzhou Industrial Park, there are various manufacturers of PC devices. Except for the CPU, which is still relying on imports, suppliers of all the other PC devices at a high quality and reasonable price are there.

2. As the PC industry has matured and the profit margin is getting thinner, the proportion of manpower cost has been more important than ever before. In Europe and the United States, there are fewer suppliers able to offer low grade products such as NC, Net PC or Basic PC, which have been well received by users on the market. Chinese manufacturers have an advantage in these products, as they have a comparative edge in manufacturing cost, marketing cost and management cost.

3. The great potential of the Chinese market provides an opportunity for local manufacturers and will certainly bring up a few large manufacturers with international influence. Local manufacturers have the advantage of building an extensive and effective marketing network and service network, and access to domestic markets makes it much easier for local manufacturers to design appropriate products according to their comprehension of the users' real demands. The local manufacturers also have an advantage in operation by avoiding long distance transportation and various complicated procedures and formalities.

The rapid growth of domestic PC manufacturers proves this analysis. What is more encouraging for PC manufacturers is the high speed growth of the PC market. In 1997, the total sales of PCs in China came up 2.25 million, and it is expected that by the year 2000 the sales of PCs will come up to about 10 million. China's PC manufacturers are greatly encouraged by the prospects. They are trying to increase their financial input and develop more technological innovation for the domestic computer industry.

China's effort in developing the computer industry also covers the development of high performance computers, which is much more complicated than the PC industry, as the former is more market oriented and the latter is more oriented to technological innovation. The successes of Yinghe II (Milkway II) and Shuguang high performance computer have shown the breakthrough that China's scientific community has made in the development of high performance computers.

China did pass through an uneven path in developing high performance computers. It was not until recent years, with the success of a host of supercomputer systems, including Shuguang 1000 and Yinghe II, that the development of high performance computer had overcome its difficulties; eventually, it became the dawn of the high performance computer industry in China.

Having tackled the key issues for a few years, the Shuguang high performance computer has formed a product series and stepped into a normal industrial cycle. The Shuguang 1000 large scale parallel computer system was a landmark for the development of the high performance computer, while the newly developed product of Shuguang series, the Shuguang Tianchao supercomputer system, is a further step in the development of the high performance computer. Shuguang, jointly developed by the State IA Computer Research and Development Center and Shuguang Information Industrial Ltd., demonstrates the industrialization of 863 high technology research results.

Shuguang has merged the international mainstream technology to support international standards and kept its unique innovation, which makes available its extraordinary high performance and broad scope of application. Only a few large computer manufacturers in the world are able to produce such high performance computer systems. The ratio of performance to price of Shuguang is double that of similar products.

The Shuguang supercomputer system adopts distributed processing memory. Its nodes come up to one thousand, the total volume of internal memory comes up to 1TB, and the computing speed is more than 4 billion times/second. RISC chips are adopted for the nodes system and connected by Ethernet and cavity communication net. The system possesses open-ended nodes, open-ended users, open-ended promotion and open-ended application. It can support the operational systems of AIX/Solaris/Windows NT and data bases of Oracle/Sybase/DB2 as well as about 10,000 systems software and application. The graphic interface developed by Shuguang adopts Java or WWW technology. The system has a good parallel programming environment, parallel user interface, parallel analysis tools and a parallel debugging environment.

Shuguang system has been adopted by quite a few cities to build high performance computing centers and has been applied in the complex fields such as meteorology and aerospace. The success of the development of Shuguang series marks a good start for the high performance computer industry in China.

With development through more than a decade, China's software industry has initially come into shape. A number of software enterprises have been established with the expansion of the software market. In 1996, the total sales of these software companies reached 9.2 billion yuan. Of course, the size of China's software industry and its market share is not comparable to the United States or even a developing country like India. Currently, sales of China's software account for less than 1% of that of the world and less than 30% of the domestic software market. All system software and most parts of supporting software come from foreign companies. In the past couple of years, even Chinese character software, which had been a field in which China was superior, has felt the impact of international competition. To turn around China's software industry from weak to strong, the experts in the country have found that quite a few problems need to be settled.

First of all is the mentality. In China, people have been used to paying attention to hardware and looking down upon software. People didn't see that, due to the characteristics of software development, the industry needs to be well organized, under strict management and highly industrialized. They didn't take software production as an "industry" at all. Therefore, much of the software development had been conducted by universities and research institutes on a small scale and in a narrow scope. Second, there was a tendency to neglect market demand in software development. Third, large scale software industrial groups and an industrial base still hadn't been set up in the country. Finally, it is necessary to work out a whole set of financial, taxation and trade policies to support the development of China's software industry.

Currently, these problems are receiving attention. Recently, quite a few large software industrial groups have been established. For instance, in Xian, a large software group sponsored by 14 enterprises, research institutes and universities has just been established, with registered capital of 125 million yuan. The comprehensive S&T force of Xian ranks third in the country. The city has a comparative superiority in science manpower, technical equipment and R&D facilities. In recent years, rapid software development has appeared in the city, as more than 500 small software companies have been started. However, the smallness of scale has limited the development of these companies and made it difficult to form a software industry in a real sense. The formation of a strong software industrial group is a significant breakthrough in software development. In China quite a few software industrial groups have developed as in Xian. China has great potential in software manpower as well as a software market; if only the software forces can be well organized, the development of software industry will accelerate.

China has the largest potential electronic information market. According to the forecast of domestic experts, by the year 2000 China will have an electronic information market of one thousand billion yuan. By the year 2010, it will come up to 6,000 billion yuan. According to the forecast of foreign experts, China will be the world's largest electronic information market in the twenty-first century. Confronted with such a huge market, China's industrial force is comparatively weak; and, therefore, China's electronic information market has been the focus of various global corporations. In December, 1996, 39 countries, including the United States, Japan, the European Community and South Korea, signed an agreement on information technology (ITA) and declared that by the year 2000 the customs duty on information products among these countries will be reduced to zero. The total amount of trade of information products of the 39 countries accounts for 92.5% of that of the world. It is indeed a great challenge for developing countries, and it is predicted that there will be an intensified struggle among global corporations in entering and scrambling for China's electronic information market.

China's information technology industry should closely follow the change of the information industry brought about by the trend of international networking and find out opportunities for China's own development. The expansion of China's market demands makes it possible to exchange technology with the markets. This will be a good opportunity for the development of the electronic information industry in China. If China can seize the opportunity, China's own electronic information industry will be able to experience high speed growth and make it possible for China to grow into an information power as well as an economic power.

The guideline for the development of a national information infrastructure, as indicated by Zhou Jiahua, former Vice Premier of State Council, is "to have an overall plan, to put the State in the forefront, to unify the standards, to construct by joint efforts, to carry out mutual connection, and to share common resources."

Through years of construction, the telecommunications networks, broadcasting television networks and computer networks have formed into an important infrastructure of China's information construction and developed to a certain scale. The telecommunications network is developing along the direction of integrated service digital network (ISDN), the broadcasting television network along the direction of multiple-functions and two-way transmission, and the computer network along the direction of multimedia and interactive systems. It is expected that the three networks can be connected and coordinated for the construction of a powerful national information infrastructure with broad coverage.

THE PETROCHEMICAL INDUSTRY

The history of the development of the petrochemical industry in China is relatively short, but its speed of growth is relatively high. The industry has come up to a certain scale, and the chief oil refinery and petrochemical technologies are available. There has been rapid development in the past decade.

The focus of the development of the petrochemical industry is to coordinate and optimize the development of the refinery industry and chemical industry. Of course, furthering refinery capacity is the key.

In order to develop the refinery industry, efforts have been made to

Table 20
Growth of the Chemical Fertilizer Industry, 1992–1996 (Unit: 10,000 Tons)

1992	2,099
1993	2,016
1994	2,276
1995	2,450
1996	2,660

Source: The Statistical Bulletin on the National Economic and Social Development of the PRC by the State Statistical Bureau.

tap the potential and carry out technological renovation. Emphasis has been placed on secondary refinery and the improvement of the capability of storage and transportation. In addition, in order to accelerate the extension of refinery technology into chemical processing, when continuing the construction of ethylene production facilities, catalytic cracking technology has been developed in the refinery industry. Heavy fraction between 346 and 546C could be used as raw material in applying catalytic cracking technology. This is appropriate for the production and application of domestic oil, as there is less light fraction crude oil in the country.

The development of chemical industry focuses on chemical fertilizer and chemical fiber (Table 20). The chemical fertilizer industry has had rapid development in the past decade, although there was the problem of repeated imports of chemical fertilizer equipment in whole sets.

The chemical fiber industry is of special importance in China, as China's textile industry is now more than ever dependent on the supply of chemical fibers. The textile industry is a conventional industry in China with a long tradition, and textile products are traditionally export products. In 1994, textile products for export accounted for 29.4% of China's total exports. China's synthetic fiber industry is dependent on the textile industry, as an overwhelming majority of synthetic fibers are used by the textile industry and very few of them are for direct export. In turn, the dependence of textile industry on synthetic fibers is more important than ever before. In 1994, 7.80 million tons of fibers were needed by the textile industry, of which polyester, propane nitrile and chinlon reached 2.82 million tons, accounting for 36% of the total.

China's synthetic fiber industry started in the late 1950s and total output came to 30,000 tons in 1970. In 1985, the total output of three major synthetic materials, plastics, synthetic fibers and synthetic rubber, was only 2.18 million tons of the total and importing of the three synthetic materials was 2.70 million tons. In the 1990s, the synthetic fiber industry saw sustained growth (Table 21).

Table 21
Growth of Synthetic Fibers Production, 1992–1995 (Unit: 10,000 Tons)

1992	208.3
1993	221.5
1994	269.0
1995	290.0

Source: The Statistical Bulletin on the National Economic and Social Development of the PRC by the State Statistical Bureau.

Still, the output of synthetic fibers was unable to meet the demands of the textile industry. In 1994, there was a total import of one million tons of synthetic fibers, including 60% polyester and 35% acrylic fibers.

By the year 2000, the demands for synthetic fibers will be approximately 4 million tons/year, and the demands for raw materials for the synthetic fibers will have a corresponding increase. For instance, the demand for polyester, PTA and ethylene and their monomers will be 2.51 million tons, 2.19 million tons and 1.06 million tons respectively.

According to the Outline, the capacity of raw materials of synthetic fibers will increase to about 4.80 million tons/year by 2000. The development of polyester will be emphasized in the development of synthetic fibers, and the proportion of chinlon and acrylic fibers will proportionately increase. The output of polyester had increased from 1.042 million tons in 1990 to 1.515 tons in 1994. It is estimated that by the year 2000, the output of polyester will increase to 2.40 million ton/year. The output of chinlon and acrylic fibers will increase proportionately to adjust for the proportion of the synthetic fibers mix.

In the process of developing a petrochemical industry into one of the mainstay industries in China, there is a major issue to deal with. Emphasis should be placed on scale operation and the scale effect of the development of petrochemical industry. For instance, in the introduction of acrylonitrile production equipment, the investment per unit product of a whole set of equipment at the capacity of 25,000 tons is 145% that for a capacity of 50,000 tons. In view of this, criteria on the scale of chemical industry equipment to be introduced was included in the State criteria for the economic scale of fixed asset investment projects, which was an appendix of the State Industrial Policy in the Nineties issued in 1994 (Table 22).

These few examples adopted from the State criteria for the economic scale of fixed asset investment projects show the determination of the government to carry out scale operations in the petrochemical industry.

Table 22
Criteria of Economic Scale of Fixed Asset Investment Projects (Petrochemical Industry)

Product Name	Criteria of Scale (Production Capacity/Year)	Illustration
Polyethylene	Newly Developed Project: Above 300,000 Tons	Aside from the ethylene projects listed in State ten year plan, no new projects will be reviewed and approved during the eighth five year plan; no new ethylene projects below the scale of 300,000 tons will be approved thereafter.
Ethylene Glycol	Newly Developed Projects: Above 100,000 Tons	
Polyester Used for Fibers	Newly Developed Projects: Single Series Production Equipment at the Scale of 60,000 Tons or Above	

Source: "Outline of the State Industrial Policy in the Nineties"

THE AUTOMOBILE INDUSTRY

The development of automobile industry, of course, covers various types of motor vehicles, but at the core is the development of the sedan car. The automobile industry occupies an important position in the national economy. With the development of a national economy and the improvement of people's living standard, the social demand for the automobile has grown rapidly, and conflict between supply and demand has become more intensified. Up to 1988, China's annual output of sedan cars was only 30,000 (including 20,000 Jeeps), which was far behind the social demand.

The automobile industry has been the only industry in the world with high industrial output, a broad scope of interconnection and a long production chain. Therefore, it can play a leading role in economic and social development. It is quite possible that China's automobile industry will become an important leader in the industrial sector and will bring along the development of metallurgical, chemical, petroleum, machinery, rubber and electronic industries as well as the development of the road system and city construction and tertiary industry. This will form an important component of economic growth.

China's automobile industry was first established in the early 1950s with support from the Soviet Union. Changchun First Automobile Factory was the cradle of China's automobile industry. The light truck had long been the chief product of the automobile industry in China. The sedan car made up an insignificant proportion of the industry. By 1995, the total automobile output increased to 1,500,000 per year, but the output of the sedan car was only 250,000. The development of sedan car production is and will be the core of the growth of China's automobile industry.

The automobile industry has a history of more than 40 years since its birth in China, but up until now it is still an immature industry. Of course, the development of the automobile industry was restricted by the planned economy for a long time. Currently, obstacles mainly come from the dispersed investment and the small economic scale that blocked further development of the industry. In China, there are 112 automobile manufacturing enterprises excluding refitting and re-equipping factories, which are in the thousands. However, the total output of the 112 automobile manufacturing enterprises is less than that of General Motors. Therefore, it is imperative to remove barriers and promote the automobile industry to develop into a modernized manufacturing industry following the specific way of scale economy. In this regard a few key issues have to be addressed.

The outstanding characteristic of the automobile industry is volume production. Volume production has a series of characteristics, such as specialization, automation, technology intensive operations and labor intensive operations. All these characteristics call for higher capital input and form a technology barrier that is difficult for many developing countries to surmount. Therefore, in developing the automobile industry, especially the sedan car industry, China must concentrate substantial capital to scale up production. It needs to carry out large scale technology renovations in existing enterprises, develop new processing technologies (such as punching press, welding, painting and assembly) and acquire the necessary technology from abroad and develop R&D on some key issues. It is beyond the capabilities of the existing truck manufacturing industry to transform either financially or technologically into sedan car manufacturing industry. With the efforts of the past few decades, since the early 1950s China has formed an automobile industrial base that has taken medium-sized trucks as its main product line. However, it possesses a lower technological level with outdated equipment. Moreover, in comparison to the production of the sedan car, truck production is at a lower grade in both volume manufacturing and technological levels. That's why the current automobile production base in China is unable to directly transform into sedan car production.

Though China's machinery industry is a long line industry, the equipment is relatively obsolete, with fewer precision processes. Therefore, there are difficulties in manufacturing special devices required by the sedan car industry. Also, the quality and precision as well as the volume production of standard parts and basic parts and batch manufacturing of China's machinery industry cannot meet the requirements of the sedan car production. Above all, the current machinery manufacturing industry can support sedan car production only in a limited way. Only with proper technological renovation as well as technology acquisition from abroad can it support sedan car manufacturing.

It is necessary to face both domestic and international markets and develop the sedan car industry at a high starting point. Currently, an organized world framework of sedan car manufacturing has been formed. All large automobile corporations in the world aim for the global market to carry out their production and marketing operation. China will be under the pressure and competition of international monopolized groups for a long period of time. China can use the domestic market as support and the international market as a goal, taking advantage of follow-up strategy and directly adopt international updated advanced technology, to gradually set up a sedan car manufacturing industry with international competitiveness.

It is necessary to encourage the sedan car parts and components industries to carry out relatively independent and properly led development. The parts and components industry is the basis of sedan car manufacturing. In comparison with the production of complete units, the requirement for the manufacturing technology of parts and components is not that high, and there is a comparatively broader market for parts and components, which is within reach of developing countries at the present technical economic level. The economic scale of production of parts and components is generally larger than that of a complete car. Therefore, before China's sedan car manufacturing industry rises to industrial scale, the parts and components production should enter the international markets to ensure that necessary market demands satisfy and support the development of a parts and components industry. Only by forming a parts and components production system with both export and import capability can the sedan car industry develop smoothly.

It is necessary to acquire advanced technology from abroad and strengthen self-development. During the early stage of development of the sedan car industry in China, it is necessary to import at a large scale advanced manufacturing technology and management technology from abroad to form productive capability rapidly. However, when introducing foreign technology, attention must be paid to strengthening the capability of self-development, especially forming the capability to develop and design a complete car. Only by so doing can China rid itself of control by global corporations and develop China's own independent sedan car manufacturing industry.

It is imperative to adhere to the principle of maintaining independence by keeping initiatives in China's own hands, taking foreign indirect investment as main goal and adopting multiple forms and multiple channels for technology acquisition. In order to develop China's sedan car industry at a high starting point, it is necessary to prepare to actively import technology and capital. However, it is also necessary to make specific analyses and deal with different cases in different ways when importing foreign capital. When it is possible to return the principal and interest, investment should be actively sought from international financial institutions. Care should be taken with joint ventures. Experience in China and many developing countries has proved that the international automobile monopolized groups use joint ventures merely as a means to market products. Only in circumstances of joint investment, joint product marketing and joint development of new products can the joint venture become a cooperative form with equality and mutual benefits in favor of the development of sedan car manufacturing industry in one's own country.

With regard to selecting the types of sedan car for development, the common sedan car with engine displacement of 1–2 liters should be developed first. The sedan car of this grade accounts for more than 70% of total sedan car output worldwide and, therefore, has larger international market demand. To give priority to the development of this type of sedan car also conforms to the basic national conditions in China, including purchasing power, the state of the road system, technological levels, and so on.

The large and sustained demand for automobiles, especially for sedan cars, provides sound market conditions for the development of the automobile industry in China. Through construction in the past few decades, China has formed the initial basis for the automobile industry. Through proper renovation, China will be able to transform it into a modernized industry and develop especially a modernized sedan car industry. The time is now for China's automobile industry to take off.

In view of this, the State Planning Commission, for the first time in 1994, issued an industrial policy for automobile industry. In general, the industrial policy was formulated with a view to building China's automobile industry (including the motorcycle industry) into a mainstay industry of the national economy as fast as possible; to turning around the current state of dispersed investment, small production scale and outdated products of the automobile industry; to strengthening the development capability of automobile manufacturing enterprises and improving the quality of products and level of technological equipment and to promoting the rationalization of industrial organization and realizing scale economy.

Policy Goals and Focal Points of Product Development

The State is to direct automobile industrial enterprises to fully utilize capital both at home and abroad and strive to exploit both domestic and international markets and carry out industrial development by adopting large volume production mode with multiple varieties of products. By 2000, the total automobile output should meet more than 90% of domestic market demand, and the sedan car output should reach more than half of the total automobile output and basically satisfy the family demand for the sedan car; motorcycle output should meet the domestic demands basically and with parts for export.

The State is to promote concentrating investment in the automobile industry and the reorganization of the industry. Focuses shall be on correcting the excessive number of production factories and the dispersion of investment, the chaotic situation of issuing project approvals, the repetition of importing low level products, and the slowness of constructing special designated factories and the process of localization. The targets shall be staged. During the eighth five year plan, the focus will be to support the completion of projects of complete car and parts and components manufacturing approved by the State and put into production, so as to prepare for the accelerated development of automobile industry at the next step. Within this century, 2–3 automobile manufacturing enterprises (groups) shall be supported to quickly grow into large scale enterprises with considerable strength and 6–7 automobile manufacturing enterprises shall be supported to grow into domestic backbone enterprises and 8–10 motorcycle manufacturing enterprises shall be supported to grow into key enterprises within both domestic and international markets. It is necessary to form a system with fewer production factories and large volume production as well as to form a market with orderly competition among a small number of large enterprises and to bring the sales of enterprises that rank among the top three in product output in each category (to be classified according to QC/T559–93 sector standard) to comprise more than 70% of the domestic market share. In the meantime, it is necessary to direct large scale enterprises and backbone enterprises to carry out "mergers between the giants" to form 3–4 large scale enterprise groups with certain international competitiveness and 3–4 large scale motorcycle manufacturing enterprise groups, so as to realize independent development, independent production and independent marketing and to participate in international competition.

Focal points of product development:

1. Automobile parts and components and key sedan car parts and components;
2. Passenger car (Category M),[1] economic sedan car, the special underframe of large-sized and medium-sized buses;
3. Truck (Category N): trucks in series, new type engine;
4. Motorcycle (Category L): engine;
5. Technological equipment: die;
6. Basic parts: forge and cast blank pieces.

Industrial Organization Policy

The purpose of readjusting the automobile industrial organization is to promote the grouping of enterprises, the development of products in series and specialized production; to effectively utilize the existing automobile industrial base and to fully mobilize the initiatives of the central departments, localities and enterprises; and to avoid blind competition at low efficiency and optimize the organizational structure of the industry.

The State encourages the automobile industrial enterprises, through various forms such as annexing assets, merging and carrying out a stockholding system, to develop transdepartmental, transregional enterprise groups, and, in connection with the reform of the property rights system of State-owned enterprises, to accelerate the transformation to a modern corporation, so as to build a modern enterprise system in the automobile industry.

The State will focus on supporting manufacturing enterprises and groups at a certain production scale that will produce automobiles, motorcycles, or parts and components with independent products, technology development capability and a certain market share. The following are the conditions and development goals that the enterprises and enterprise groups supported by the State should make available before the end of 1995:

1. For those with a production scale at an annual output of more than 300,000 automobiles, annual sales of more than 200,000 automobiles and technology development expenditure not less than 3% of sales, the State will support them in reaching their development goal of an annual output of 600,000 automobiles.

2. For those with a production scale at an annual output of more than 150,000 automobiles, annual sales more than 100,000 automobiles and technology development expenditure not less than 2.5% of sales, the State will support them in reaching their development goal of an annual output of 300,000 automobiles.

3. For those with a production scale at an annual output of more than 100,000 automobiles, annual sales more than 80,000 automobiles and

technology development expenditure not less than 2% of sales, the State will support them in reaching their development goal of an annual output of more than 200,000 automobiles.

4. For those with a production scale at an annual output of 20,000 heavy duty automobiles, annual sales more than 10,000 automobiles and technology development expenditure not less than 2% of sales, the State will support them in renewing their products and reaching an appropriate scale.

5. For those with a production scale at an annual output of 1,500 large and medium-sized passenger cars or underframes of passenger cars, annual sales more than 1,000 cars and technology development expenditure not less than 2% of sales, the State will support them to reach an appropriate development goal.

6. For those which keep more than 25% of domestic market share in the same category of products with the key parts and components of the sedan car they manufacture or fill the void in a field that is to be urgently developed (the catalogue of such fields to be identified), the State will support them to reach their development goal of scale economy.

7. For those which keep more than 10% of market share with motorcycles they manufacture, the State will support them to further expand their production capacity and increase varieties.

The calculation of production scale and sales of the enterprises and enterprise groups includes those of the parent company and its wholly owned subsidiaries, holding companies and China-foreign joint venture subsidiaries.

Enterprises that conform to the aforesaid stipulations of the industrial policy, starting in 1996, when developing products included in the focal points of product development, either by new construction, renovation or expansion, through State approval, may enjoy the following benefits:

1. A tax adjustment will be made on fixed assets investment with a zero tax rate.
2. The stocks and bonds of these enterprises will be arranged in priority for distribution and going public.
3. Active support will be offered in obtaining bank loans.
4. These enterprises may be given priority in using overseas capital.
5. Policy debts will be appropriately arranged for projects of the sedan car, key parts and components of the sedan car, die, forge and cast pieces.
6. The financial companies under the enterprise groups, through State approval, may expand their scope of business operation.

Projects producing the complete car and engine newly approved by the State (including China-foreign equitable and contractual joint venture projects) will be constructed according to the following output scales:

1. For the sedan car projects with an engine displacement below 1600 cc, an annual output not less than 150,000.
2. For light truck projects, an annual output not less than 100,000.
3. For light passenger car projects, an annual output not less than 50,000.
4. For heavy truck projects, an annual output not less than 10,000.
5. For motorcycle projects with an engine displacement below 150 cc, an annual output not less than 200,000.
6. For gasoline engine automobile projects with an engine displacement below 2500 cc, an annual output not less than 15,000.
7. For diesel engine automobile projects with an engine displacement below 2500 cc, an annual output not less than 100,000.

Industrial Technology Policy

The State encourages and supports automobile manufacturing enterprises to establish their own product development and R&D organizations to assimilate and absorb foreign technology, so as to form an independent product development capability. For the major R&D projects jointly developed by different enterprise groups, the State will provide support by offering research and development funds.

The State encourages the extension and use of electronic automobile technology and new techniques, new materials and the production of energy saving and low pollution automobile products, and research and development on automobiles that use new fuels and power.

The newly constructed automobile manufacturing enterprises should ensure the advancement of their products. The renovation of existing products and newly developed products should be brought up to international standards during the early 1990s, and the products of imported technology should reach the same international standards.

Passenger cars and trucks with a total mass of no more than 3.55 tons should gradually adopt gasoline above No. 90 as fuel before the year 2000; passenger cars with a total mass of no more than 2 tons should completely adopt lead-free gasoline; passenger cars and trucks with a total mass of more than 5 tons should mainly adopt diesel oil as fuel after 2000.

The State supports setting up State level automobile, motorcycle and key parts and components research, experiment, examination and test organizations to undertake the task of formulating standards, product authentication, examination and verification of import and export goods.

The State encourages the automobile manufacturing enterprises to establish technology research and development companies in the form of either equitable or contractual joint ventures with foreign enterprises.

The State encourages the automobile manufacturing enterprises to adopt modern electronic technology, flexible processing equipment and on-line automatic checkout and test equipment and to select and use automatic equipment with specific targets, so as to enhance the per capita equipment rate and the technological level of equipment.

Policy on Using Foreign Capital

The State encourages automobile manufacturing enterprises to use foreign capital to develop China's automobile industry.

When an automobile manufacturing enterprise uses foreign capital directly, it should select a foreign enterprise that conforms to all the following conditions as an equitable or contractual partner:

1. Possesses its own product patent rights and trademark rights.
2. Possesses its own product development technology and manufacturing technology, and the technical indexes of its product conform to the decrees of the country (or region) in which it is located.
3. Possesses independent marketing channels.
4. Possesses sufficient financing capability.

The foreign enterprise shall never set up two or more equitable or contractual joint venture enterprises in China for the production of a complete car in the same category.

For automobile manufacturing enterprises with advanced product development and manufacturing technology, the State will support them in directly using overseas financial capital or in indirectly using foreign capital for development.

China-foreign equitable or contractual joint venture enterprises can be established only by satisfying the following conditions:

1. The establishment of a technology research and development organization within the enterprise that is available to the main development capability for product renewal.
2. The product is manufactured at the international technological level of the 1990s internationally.
3. The enterprise shall export products made by itself as primary and settle the foreign exchange equilibrium by itself.
4. When the joint venture enterprise is selecting and using parts and components, locally made parts and components shall have the right of first refusal.

For the China-foreign equitable or contractual joint venture enterprise that produces whole units of automobiles, motorcycles or engines, the share of the Chinese Party should not be less than 50%.

Table 23
Growth of the Automobile Industry, 1993–1995

Year	Output of Automobiles (Unit: in 10,000s)	Increase over Previous Year (%)
1993	131.0	22.8
1994	140.2	8.0
1995	150.3	10.0

Source: The Statistical Bulletin on the National Economic and Social Development of the PRC by the State Statistical Bureau.

It is necessary to stop examination and approval of projects engaged in the operation, through whatever business forms, of renovating or scaling down the importation of used automobiles and motorcycles. The contracts of approved projects shall never be extended, and a strict supervision system should be set up to ensure that all the renovated automobiles and motorcycles and knockdown parts and components are sold abroad.

Policy on Localization[2]

The automobile manufacturing enterprises, after importing product manufacturing technology, must carry out product localization. The progress of localizing the product made by imported technology will be one of the conditions for winning support of the State for the development of a second type of automobile.

The automobile manufacturing enterprises should never use methods of SKD (semi-knockdown) and CKD (complete knockdown) to import parts and components for assembly.

In accordance with the localization rate of the automobile industry, the State has formulated a preferential tax rate on import customs duty.

The preceding points are the main contents of China's industrial policy on the development of automobile industry. In the past few years, significant progress has taken place in the industry (Table 23).

According to the figures provided by the State Statistical Bureau, the output of the sedan car in 1996 was separate from automobile. In 1996, automobile output was 1,490,000, and sedan car output was 390,000.

Currently, there are three major automobile industrial bases in China that are supported by the State. Shanghai has developed two major joint venture projects, one with Germany, which has produced a brand name sedan car—Santana—for more than a decade and another the newly established joint venture with General Motors. The output of the Santana reached 150,000 in 1997. With the two joint ventures, Shanghai is expected to rise to an annual output of 300,000 sedan cars in the coming years. The other two large automobile manufacturing bases are located in Changchun and Wuhan—the number one and number two automobile factories. The former has established a joint venture with Germany and produced a brand name sedan car in China—the Audi.

THE CONSTRUCTION INDUSTRY

During the ninth five year plan in China, the construction industry has been identified as one of China's mainstay industries and the first economic growth point of the country.

The construction industry, especially the home-building industry, has a broad market in the country. Considering the large population and the current housing situation in China, the market potential of home building is enormous. It is a general demand of people for improving their living conditions. The State has set up a standard that by 2000, the average living space per capita in cities and towns shall be 9 square meters. However, according to statistics, in 1995, in large and medium-sized cities in 42.4% of families each member occupied a space less than 10 square meters and in 18.6% of families each member occupied a space less than 6 square meters. There were 4 million families in which each member occupied a space less than 4 square meters. In 1995, the average per capita living space in China was 7.9 square meters. As it is expected that the average per capita living space will increase to 9 square meters by 2000, it is necessary to add 400 million square meters of construction area per year. As per capita living space should reach 13.7 square meters by 2010, it is necessary to add 600–800 million square meters of construction area per year. From the figures, the general demand for home building indeed appears enormous in China. According to the Ninth Five Year Plan for National Economic and Social Development and the Outline of the Long Range Goal up to 2010 of the PRC, a total construction area of a billion square meters will need to be added by 2010.

The development of home-building construction will create enormous social demands and thus bring along the development of a number of industries. First of all, a series of material industries, such as steel, lumber, cement and glass, will have an ever bigger market. People who move into their new homes need to decorate their houses, which will promote the development of a number of related light industries. Promoting the construction industry into a mainstay industry is also based on its potential to increase the prosperity of a number of industries.

Nonetheless, the development of the construction industry requires drastic reform of the present housing system. China's housing system still mainly follows the planned economy. Under the planned economy, housing was part of workers' and government functionaries' welfare support. The houses were distributed according to people's life service, status and family size. Houses were owned by the State, and dwellers paid an insignificant amount of rent. It has long been a heavy financial burden for the State to build more new houses. That's the major reason for maintaining very limited living space for people. If the government tried to expand people's living space in the traditional way, the investment would be far beyond the financial power of the State. The only way is to carry out the commercialization of the housing system. A house is a commodity in a market economy. It is normal in a market economy to develop a real estate market in which people buy their houses and to make it possible for people to transfer the property rights of their houses. The prosperity of the real estate market will not be a burden on the State; on the contrary, it will bring in a lot of revenue for the State. In the past decade, there has been rapid growth in people's personal savings. In 1996, total personal bank savings came to 3852.1 billion yuan. Home purchases will be a good way to divert people's consumption. Since the early 1990s, the real estate market had developed rapidly, but it mainly focused on the development of luxury buildings, such as high standard office buildings for foreign businesses, luxury apartments for foreign businessmen, luxury hotels and shopping centers and villas for small groups of wealthy individual businessmen. Now, the high standard real estate market is saturated. For instance, the vacancy rate of luxury hotels has risen except during the high tourist season, the price of high standard office buildings has dropped and sales of luxury shopping centers have been at a standstill or fallen. According to statistics, right now, the total space of luxury shopping centers in Beijing is four times that in New York City, while the purchasing power in Beijing is only one-eighth that of New York City. A few years later, when all the new shopping center construction projects are completed, the total space of luxury shopping centers in Beijing will be eight times that of New York City. Therefore, the high grade real estate market is a limited investment field with a limited source of customers. However, personal home building will have a huge potential market and will be supported by people's strong desire to expand their living space and to own their own homes and is supported by the astronomical figure of personal bank savings. Choosing the construction industry as a mainstay industry is indeed a good choice, and the emphasis on home-building development is the right emphasis.

The development of the home-building market is only now starting, and further development of the market will require overcoming quite a few difficulties. The astronomical figure of personal savings came from the improvement in people's income and was due to the large population. There is still a big gap between house prices and people's average income. People's average income is 5,000 yuan per year, while the average house price is 1,500 yuan per square meter. In a few large cities, the price is as high as 5,000 per square meter in the main part of town; and in individual places, it is as high as 10,000 yuan per square meter. According to experiences in foreign countries, it is only when the house price is 3–6 times personal income that it is possible for people to buy houses. Now, the house price is 16–18 times people's average income in China. How can people get the money to buy houses? Therefore, the government is trying to bring down the cost of home building, including cutting 48 items of unreasonable charges on home-building construction. This has helped to lower the price of a house by 15%. In the meantime, banks have started to offer home purchasing loans and arrange for down payments for home buyers. All these measures have helped to raise the purchasing capability of people.

If the development of home-building construction is mainly due to the reformation of the housing system and market promotion, then, the development of the building material industry is mainly due to technological renovation. Previously, high standard building construction used not only domestic materials but also a large number of imported materials. For home building construction, all the materials come from domestic industries. Therefore, improving the quality and performances of domestic building materials will be of special importance.

There is urgency to renovating China's building material industry. China's building materials have been at a standstill for years. For instance, solid clay bricks and tiles have been used for wall and roof building for generations, but the making of solid clay bricks causes a high consumption of clay and energy. In 1987, the output of solid clay bricks was 379.6 billion pieces. As a result of taking earth from farmland, 50,000 mu of farmland was destroyed during that year. The average energy consumed was 1.23 tons of coal per 10,000 bricks. In 1986, the consumption of standard coal for baking bricks reached 50 million tons, accounting for 50% of the energy consumed in the building materials industry. For years it was encouraged to replace solid clay brick with cavity brick and use various industrial waste residue including powered coal ashes and coal gangue to replace clay in making bricks. In order to increase the proportion of cavity bricks, the tax rate on cavity bricks has been reduced.

With respect to cement production, energy saving is the focus. A new cement dry processing technology with the pre-decomposition kiln at its core has been developed and extensively used in the cement industry in order to reduce energy consumption by a big margin. Efforts have been made in promoting the adoption of bulk cement.

According to the Ninth Five Year Plan for National Economic and Social Development and the Outline of the Long Range Goal up to 2010 of the PRC, in the coming five years, the building material industry shall focus on readjusting its industrial formation, energy savings, land conservation, water conservation and reducing environmental pollution. It is necessary to accelerate the reengineering and expansion of old enterprises and to enhance the level of the scale of the production line unit to endeavor to increase products of fine quality and develop and extend building materials and new types of products. It is necessary to enhance the bulk cement rate, develop commercial concrete, restrict production and use of solid clay bricks and promote the use of industrial waste residue in producing wall materials. It is also necessary to promote excavation and processing of non-metal minerals and develop inorganic non-metal materials. By 2000, the proportion of rotary kiln output shall account for 25%.

The plan for home-building construction is closely related to city planning as well as the development of the city infrastructure. Now, the newly developed city home buildings are formed in a number of small comprehensive residential areas. Each small comprehensive residential area, centered around homes, has various service facilities available, such as central heating facilities, primary or secondary schools, supermarkets, green space, recreation facilities and so on. The development of a large number of small residential areas certainly requires the expansion of the road system, the city transportation system, water, electricity and gas supply systems, sewage and drainage systems and the like. The construction cost of home building has covered part of the expense of infrastructure development. Therefore, large scale home-building construction will bring along the city infrastructure development and provide impetus for economic growth as well as increased employment opportunities.

NOTES

1. According to the State criteria of classification of motor vehicles (QC/T59-93) issued by State Technology Supervision Bureau, M category stands for passenger car, N category stands for truck and L category stands for motorcycle.
2. According to the definition given by the State Industrial Policy of Automobile Industry, localization covers everything produced within the territory of the PRC.

15

Emphasis on Basic Research

In the 1980s, the status of basic research was a controversial issue in the scientific community. The overall strategy of science policy in the 1980s was to adopt a follow-up strategy on the development of basic research. In the 1990s, when readjusting the science policy, the follow-up strategy was replaced by a strategy of scaling new heights in science. This is a big change.

In the process of readjusting the science policy, discussion has been carried out on the importance of basic research as well as how to fit basic research with the goals of the State. The fundamental objective of basic research is to explore the essence of Nature and discover the changes and laws in the evolution of Nature, so as to broaden the scope and further the depth of people's understanding of Nature. This is a general approach to basic research. What merits special attention is the relation of basic research to national economic development. It is a realistic approach.

During the readjustment of China's S&T policy, the importance given basic research called attention to the scientific community. Li Zhengdao, the world-famous physicist and Nobel prize winner, put forward an interesting metaphor, which has been influential in China's scientific community. He described the relationship between basic research, applied research and development as similar to the relationship between water, fish and the fish market. He said: "Fish can not survive without water, and there won't be any fish market without fish. So, when people try to develop a fish market, they should, of course, prepare for more fish and, at the same time, a sufficient amount of water should never be neglected."[1]

Regarding the State goals of research, experts placed emphasis on integrating near-term interests with long-term interests. This means that the goal of basic research should conform to the national conditions in contemporary China, a country with a large population and a shortage of resources; at the same time, it should adapt to China's demands for gradual development into a world power, including a world scientific and technological power in the coming 50 years.

When setting up the goal of basic research, people should take a practical approach. However, in conducting basic research, it is necessary to insist on high quality. Regarding the quality of research, it is indicated by the relevant government authority that in contemporary S&T development, whoever develops more innovations, achieves a higher level and makes bigger breakthroughs will take the initiative and seize the opportunity to catch up from behind. Only by making breakthroughs in scientific results and raising high level talent can a country scale new heights. Basic research is key to enhancing the capability of technology innovation and development on China's own accord.

In the contemporary world, basic research is not only the personal interest or personal behavior of scientists but also a scientific cause with certain social benefits. It has been a trend in the development of science and technology in the contemporary world that major basic research tasks are organized and promoted by the government or scientific institutions directly. Therefore, when using channels, including the science foundation, to support the self-selected topics put forward by scientists according to their personal specialty and intention or the key projects selected by scientists according to the requirements of disciplinary development in priority fields, it is also necessary to organize and carry out basic research initiatives on major key projects acknowledged by the scientific and technological community as the topics of overall and lead importance to national development as well as to progress in science and technology.

According to this principle, a series of basic research and applied basic research fields were selected by the State Mid-term and Long-term Scientific and Technological Development Outline,[2] including:

1. Strengthening research on the basic disciplines at the frontier, including condensed state physics, molecular biology, chemistry, oceanography, ecology and information science, especially to make breakthroughs in the physical mechanism of high critical temperature superconductors, strengthen the research on interdisciplinary fields between physics and life science, material science and energy science and

strengthen research on non-linear mathematics, computing mathematics and the theory and method of large scale science and engineering computation.

2. Molecular and cell biology research. It is essential to explore the basic laws of genetics, differentiation and development on the level of molecule and cell, so as to provide modern biotechnology development with a theoretical basis; to strengthen chemical research related to the biological process; to develop macromolecule, catalyst, surface and rare earth chemistry centered around the exploration of new materials; and to pay attention to research and the tracking of molecular design and micro reaction kinetics.

3. In respect to earth sciences, it is necessary to conduct research on the complicated coupling system composed by solid earth, atmosphere, hydrosphere and the entire biosphere, so as to provide basic materials and a theoretical basis for solving a series of major issues including resources, energy, environment and natural disasters of the country. Ecological research should stress the mechanism of the system of coordinated evolution and degeneration of the ecological system and the optimization of the composition of artificial systems, so as to promote social development.

4. In research on information science, emphasis should be placed on developing artificial intelligence to solve the problems of information technology, such as telecommunications, computers and automation control, and to strive for major breakthroughs.

5. For the development of space science, it is necessary to strengthen research on basic space science, closely integrate it with high technology research and stress physical, chemical and biological phenomena in microgravity and the integral behavior of the solar-terrestrial system.

In order to support major basic research projects on the scientific frontier with an importance of application and maximize the superiority of resources and scientific talents in China, the Scaling New Heights Program began in 1992. The Scaling New Heights Program is a basic research program consisting of two parts. The first part covers projects to be carried out in a period of five years beginning in 1992. The second part includes 32 tentative projects to be carried out later. The criteria for the project selection are:

1. Basic research projects on the scientific frontier with a better research basis on which China may make major breakthroughs by the end of the century.

2. Basic research projects with an important prospect of application, badly needed by China's economic construction and actively pursued internationally, through which China may make excellent accomplishments by the end of the century; basic research projects that can maximize the characteristics of China's natural geography and resources and

may obtain research results with China's own characteristics within the century.

3. Basic research projects in areas in which China possesses superiority and can take the lead internationally and may make further major progress by the end of the century.

In accordance with these criteria, 12 projects were selected and carried out as the first part of the Scaling New Heights Projects:

1. Basic Research on High Critical Temperature Super-conductor Electrical Property
2. Research on Structure, Performance, Molecular Designing and the Process of Preparing Photoelectron Function Material
3. Research on Climate Kinetics and Weather Forecast Theory
4. Theory and Methods of Large Scale Science and Engineering Computations
5. Exploration of Super Lattice Physics, Material and Structure of New Device of Semiconductors
6. Basic Research on Male Sterile Hybrid Superiority of Grain, Cotton and Oil Plants
7. Forecast and Countermeasure Research on the Change of China's Future (20–25 years) Living Environment
8. Non-linear Science
9. Research on Optimized Nodulation Nitrogen Fixation Control Model in Symbiotic Nitrogen Fixation System
10. Research and Application of Modern Crust Movement and Earth Kinetics
11. Research on Channels and Colaterals[3]
12. Research on Major Chemical Issues in Biological Process[4]

In order to promote the development of basic research, it is necessary to have a highly trained qualified research force that will perform basic research. It is an important component of the newly readjusted science policy that a scientific force, limited in number but highly trained, should be kept in a small number of basic research institutes, State key laboratories, State engineering centers and public welfare institutions, with a view to having a small but highly trained research force make strides on the contemporary scientific frontier to scale new heights of science and technology. The State will give support, on a selective basis, and increase financial input to ensure the development of the scientific cause.

The State key laboratories have been set up in a number of selected universities and research institutes under the Chinese Academy of Sciences and various ministries, which have certain outstanding scientists, good research facilities, better experimental conditions and a good research basis in different scientific disciplines. For instance, a genetic engineering laboratory was set up in Fudan University with the backing of the Genetics Research Institute at Fudan University. A laser technology laboratory was set up at the Central China Science and Technology University with the backing of the Laser Research Institute of the University. Likewise, the Oceanography Laboratory was set up in Shanghai Jiaotong University, the Tribology Laboratory was set up in Qinghua University and the Protein Engineering and Plant Genetic Engineering Laboratory was set up in Beijing University. Up until now, 155 State key laboratories have been established, and 40% of these laboratories are working in basic research fields and 60% on applied basic research projects. All these laboratories have undertaken more than 5,000 State projects and more than 400 international cooperative research projects, publishing 400 papers on international scientific journals, 12,000 papers on domestic journals, and more than 600 monographs in Chinese and foreign languages. More than 4,000 master's degree holders, 1,700 Ph.D.'s and 400 post-doctorates have been brought through these laboratories. More than 700 research results have passed through State appraisal, and more than 200 of these have been patented. All these State laboratories have greatly supported the development of the major scientific research of the State.

The number of State engineering centers that provide industrial technology with basic technological support has increased to 156, and 46 of these have been through State review and put into operation. By 2000, the total number of State engineering centers will come to 200, with ranks made up of 30,000–40,000 highly trained scientists and engineers working with these centers.

Though both key laboratories and engineering research centers have strong academic backing as well as financial support from the State, it still needs to address efficiency in research work. In order to ensure higher efficiency, a competitive mechanism has been introduced into the basic research system. Basic research projects should first apply to the relevant authority for funding on a competitive basis; projects are selected through peer review. The research force may be organized on a transdepartmental basis to pool the best scientists on each project. The State key laboratories are open to foreign scientists, and international cooperation is encouraged to improve the level of research.

Recently, an exhibition was held to show the scientific accomplishments China has made in the past five years. At the exhibition, a number of significant scientific results were shown to the public. For instance, China Circulation I is China's nuclear fusion experimental device. After the State appraisal in 1995, it broke a domestic record with a 2.1 second long purse discharge test, and in June, 1997, it once again changed the record high, to 4 seconds, and became a leading level for similar devices or similar international scale. The machine proof theory named "Wu's method" was founded by Wu Wenjun, Academician of the Chinese

Academy of Sciences. Proceeding from the machine proof of geometrical theorems, Wu founded a whole set of mathematical theories on mechanization. His theory has an important application in computer graphics, machinery design, theoretical physics, and the like. Wu won, with his new accomplishment, the 1991 Mathematics Award of the Third World Academy of Sciences and the Herbrand Automation Inference Accomplishment Award in 1997. The discovery of the biological explosion on Cheng Jiang River of Yunnan during the pre-Cambrian period, 510 million years ago, is another significant accomplishment of basic research. The discovery was displayed through pictures of fossil samples and a multimedia color map of the living organisms' sea floor life. The accomplishment seriously challenges Darwin's traditional evolution theory on the uniformity and anamorphosis of the origin and evolution of life. This has led to a profound change in the understanding of the origin and evolution of life. The research has been viewed as "one of the most surprising scientific discoveries in the twentieth century."

Basic research in China is starting anew under the goal of "scaling new heights of science and technology." Right now, both basic research personnel and funding are still limited. The total science and technology budget was 0.5% of the GNP in 1995; in 1996, the percentage dropped to 0.48%. It is expected that the percentage will rise to 1.5% by 2000. However, as the State is determined to consolidate a small but highly trained basic research force and research topics are decided by careful and deliberate selection to give prominence to the most important key issues, more breakthroughs are expected in the coming years.

NOTES

1. See "An Approach to the Tendency of Scientific and Technological Development from the S & T Policy of a Few Countries," Li Zhengdao, *Science and Technology Daily*, August 23, 1997.

2. "State Mid-term and Long-term Scientific and Technological Development Outline," *People's Daily*, April 11, 1992.

3. One of the basic theories in Chinese traditional medical science. It is regarded as a network of passages, through which energy circulates and along which the acupuncture points are distributed. The project focuses on the basic theory of channels and collaterals, the analysis of its morphogenesis and physiological function and the law of diagnosis and therapy. This will not only promote Chinese traditional acupuncture and acupuncture anesthesia but also bring about a leap in the understanding of biological phenomena.

4. This research mainly relates to the structure, function and chemical synthesis method of three categories of biological macromolecules related to the essence of biological activity, namely sugars, proteins and nucleic acids, as chemical substances and their change of organism are the basis of biological phenomena.

16

High-Tech Research

China's science and technology policy has undergone change in the 1980s and 1990s. In the 1980s, according to China's actual circumstances, the first priority of scientific and technological development was to remold conventional industry, which asserted a decisive impact on China's industry and national economy. Modernizing industry as well as realizing the goal of quadrupling China's total industrial and agricultural output was determined mainly by technology renovation of conventional industry. In the 1990s, when readjusting science and technology policy, greater emphasis was put on the development of high technology. The rapid growth of China's national economy and severe international competition have played a role in forcing China to take a different outlook on the development of high technology.

China started her high-tech research as a State Scientific Program in the mid-1980s. Since the mid-1970s, the development of a new round of the industrial revolution has strongly impacted the economic structure, production organization and even the social structure of developed countries. The development and application of emergent technology has brought about a leap forward in productive forces, changed industrial structures, and led to the enhancement of productivity and tremendous changes in working methods. The rise of high-tech has promoted changes in management methods and further intensified competition in employment opportunities and world markets. In the 1980s, many countries began to see high-tech development to be an important component of state strategy. It has had profound impact on the world economically, technologically, socially, politically and militarily.

The level of development of a country's science and technology has become an important mark for measuring a country's comprehensive national power and international status. The upswing of the world's new industrial revolution has been at once a serious challenge to China and an excellent opportunity for China to develop her science and technology. As early as 1983, the government indicated that it was imperative to study the countermeasures, to grasp the opportunity and to respond to the challenge. This instruction drew attention from people in various fields of the country. A consensus was reached that China should pursue, comprehend and study the latest tendencies of the world's scientific and technological development and approach it with a connection to the modernization of China. China might directly adopt the updated scientific and technological results of the contemporary world with a view to skipping certain steps in the development of conventional technology and narrowing the gap between China and developed countries. This would quicken the pace of modernization, provided China could proceed from her own national conditions and correctly work out countermeasures after reviewing past experiences.

In March of 1986, a number of well-known experts organized to expound and improve the development strategy of high technology and help in formulating the "Outline of the High Technology Development Program" by the end of the year. The Program was formally announced and implemented in 1987. As the Program was initiated in March, 1986, it was called in short "the 863 Program."

According to the circumstances in China, the main concern during that stage in China was how to accelerate the technological remolding of the national economy, so as to make it possible to quadruple the total industrial and agricultural output value by the end of the century. Even though the development of high technology was quite important, it was impossible, for the time being, to spend a lot of scientific manpower, material and financial resources to develop high-tech research on a large scale and in an all-encompassing way. It was impossible, and also unnecessary, for China to engage in intensified competition with a number of countries in the world to scramble for superiority of high technology. However, in view of the gigantic influence of high technology to future economic and social development, it was necessary to take powerful policy measures and make long-range arrangements for developing high technology in China.

Due to the reasons mentioned above, the guideline for developing high-tech research in China was limiting targets and giving prominence to the focal points. The objectives were:

1. To actively follow the international trend of developing high technology, narrowing the gap between China and the advanced level abroad and strive for breakthroughs in the few projects in which China has the superiority;
2. To train and develop a group of talented high-level scientists and engineers of the new generation with high creativity and strong international competitiveness distributed throughout a whole set of disciplines with reasonable disposition;
3. To apply the results of developmental research in national economic construction quickly and serve to renovate conventional industry and establish new industry, so as to realize modernization of industry, agriculture and service industries by the end of the century;
4. To pave the way for the formation of China's high-tech industry with certain superiority after the year 2000 and prepare reserve strength for stable and sustained development of the national economy to a higher level.

According to the above-mentioned objectives, seven specific fields were selected as focal points for development: biotechnology, space technology, information technology, laser technology, automation technology, energy technology and new materials.

As State financial resources were limited, there were limited funds to be spent on high-tech research and development. Therefore, within the selected fields, the guideline of "limiting targets and giving prominence to focal points" was also in effect. It was in this regard that scientists made careful selection of the focal points of each field.

In biotechnology, according to the actual circumstances in China, R&D focused on settling the problems in agriculture and medicine and meeting the demand to improve people's nutrition and level of health. In this regard, three topics were selected: (1) New animal and high yielding plant species with fine quality and contravariance resistance; (2) new medicines, vaccines and gene treatments and (3) protein engineering.

In space technology, R&D was focused on the advanced performance large-scale carrier rocket and a new space transportation system with improved capability of space launches and commercial service. Research on space science and space technology for peaceful purposes was to continue.

In the field of information technology, three topics were selected: (1) the intelligent computer; (2) photoelectric devices and microelectronic, photoelectric system integration technology and (3) information acquisition and processing technology.

With respect to laser technology, China has had great progress in past decades and has developed a high-level scientific force in this field. As there were broad prospects for the future application of laser technology, it was decided to include laser technology as one of the focal points of the high-tech research program, primarily through research into high efficiency and high quality laser technology to develop technological sciences with respect to pulse power technology, plasma technology, new materials and laser spectroscopy, with emphasis on the practical application of laser technology in production, processes and other fields.

In the field of automation, R&D was focused on two topics: the computer integrated manufacturing system (CIMS) and the intelligent robot. The selection of CIMS was based on the actual circumstances in China and was aimed at the demand for automated manufacturing technology for a new generation geared to the needs of manufacturing products in large varieties and small batches, with high efficiency and quality, so as to make it possible to offer a flexible reaction to the changing market and carry out fast product renewal. It was necessary to develop the key technologies of CIMS and to establish a demonstrative production line. For intelligent robot technology, it was necessary to develop intelligent robots in various types to carry out precision assembly operations, underwater operations and adverse environmental operations. This would provide the necessary technological basis for gradually formulating a robot industry in China and play a role in fundamentally changing China's industrial system and production methods.

The development of energy technology is of major importance in China. Realizing the four modernizations to a great extent depends upon the energy supply and the exploitation and effective utilization of new energy. For the time being, the chief problems were the situation of coal production and transportation, the low exploitation rate of hydropower, the shortage of petroleum and natural gas reserve and the low utilization rate of energy resources—all of these led to a shortage of power for industrial production and daily use, which had become a bottleneck in national economic development. Therefore, developing energy technology was included in the focal points of the high technology development program. In accordance with the circumstances in China, the power industry looked to thermal power primarily, and coal-fired power accounted for about 84% of thermal power. However, for the time being, coal-fired power technology was still relatively backward, coal transportation was difficult and environmental pollution was serious. So, it was necessary to carry out research and development on coal-fired magnetohydrodynamic technology to enhance the efficiency for transformation to thermal energy, to save coal resources, to use middle grade coal and high sulfur coal for power generation and to mitigate the problem of coal transportation and environmental pollution, which were derived from coal-fired power generation. The generation of nuclear power would provide an important energy resource in the next century. As part of the High Technology Development Program, in addition to the project of coal-fired power technology, nuclear power was also a focal point of energy research. The three different types of reactors are the fast neutron breeder reactor, the high temperature gas-cooled reactor (HTGR) and the fission-fusion reactor. Through pre-study and fully expounding and proving, the plan was to develop a type of reactor that might enhance the utilization rate of nuclear fuel and improve its safety and economy, so as to accelerate the exploitation of new energy resources.

In China, research and development of new materials started as early as the late 1950s and had helped to vigorously promote the development of rising industries, such as electronics, integrated circuits, new energy and space technology. However, compared to developed countries, there were still certain gaps with respect to techniques, varieties, quality and cost as well as the basis for development and technological level. Therefore, research and development of new materials was also included in the focal points of the High Technology Development Program. The focuses of R&D in this field were photoelectric information materials, high performance structural materials with corrosion resistance and low weight and special functional materials, so as to be geared to the needs of various high-tech industries.

The High Technology Development Program had both a mid- and a long-range plan. It was aimed at the frontier of the development of the world's high technology, but with limited targets. In order to make good use of limited financial resources and to gain the best effects, a series of policy measures had been taken. All these policy measures helped to direct the program to develop along the right track. After a decade of implementation, a number of results have been achieved in the selected fields.

1. In the field of biotechnology, the most outstanding achievement is the development of a sub-breed hybrid rice that is able to increase output by 10–15% over that of original hybrid rice. It is ready to be developed on a large scale during the ninth five-year plan period. Major achievement has been made in the research of genetically engineered pest resistant cotton. Outstanding results have been obtained in agricultural recombination of bacteria and artificial seeds. Breakthroughs have been made in the prevention and cure of hepatitis, while other genetically engineered vaccines have been included in international competition. Good results have been obtained in the field of gene treatment technology. The genetic engineering medicines and vaccines were developed from nothing, and now a foundation has been set for the development of high-tech industry in this field.

The High Technology Development Program has played a role in radiating and promoting biotechnology in China. What's more, the Program has given rise to a number of transcentury biologists. Though the gap between China and the developed countries in respect to the upper stream technology of biotechnology has narrowed, a larger gap remains in respect to the lower stream technology. Therefore, it is necessary to strengthen research in biotechnology and to increase the funding to further transform research results and the development of products. China has indeed reached an international level in a few biotechnological fields, but is limited in product innovation. In the future, it is essential to strengthen and encourage innovation, so as to bring China's characteristics into international competition. An urgent task is to establish research centers in biotechnology product development. In order to encourage innovation, it is still necessary to further efforts in basic research in biotechnology. Especially, special efforts need to be made in attacking the difficult key issues related to product development, in order to obtain China's own intellectual property rights and to promote the development and industrialization of biotechnology.

2. In the past decade, in the field of space technology, China has carried out a series of conceptual follow-up research projects with respect to space voyage transportation systems, large scale carrier rockets and space stations as well as space science and its application and has carried out pre-studies on selected key technologies. A number of preliminary phase results have been achieved.

Valuable research results have been achieved by using recovered satellites and final stage rockets to carry experiments on gallium arsenide materials, protein crystals, cells, microbes and plant seeds and in using balloon satellites to carry out environment probes.

With respect to research on key technologies of space to surface remote sensors, what has been carried out mainly was the expounding and proving of the scheme as well as the researching of key technologies on the medium resolution image spectrocomparator, high resolution image spectrocomparator and microwave multimodal remote sensing equipment.

With respect to experiment and research on a new generation of liquid fuel rocket engines, through conceptual research on the carrier rocket, it has been determined that the next generation carrier rocket should use a non-polluting and high performance propellant.

With respect to technology research on short-term environmental controls and life safety, breakthroughs have been made in the basic technology that uses super oxides to control cockpit pressure and gas elements. For regeneration life safety technology, research has been done on technologies of waste water treatment and regeneration, CO_2 recovery, purification and compressing. A theoretical prototype has been developed and theoretical experiments have been carried out.

Due to the limitation of financial resources, China's investment in space science is much lower than that of the world's space technology powers. China has, first of all, developed various applied satellites. Research on manned space flight and deep space probes can only be carried out within the limits of the country's resources and will gradually increase as the national economy develops. Of course, it is necessary to pursue the frontier of world space technology on a selective basis and to explore and research some key technologies that will benefit national economic construction and national defense in the next century.

3. Information technology is one of the focal points of the High Technology Development Program. The AI computer system is the major project in information technology research. In the world, there are different methods and different emphases in terms of AI research. In China, experts believe that the "AI computer" is a developmental concept. The current focus of the research work is to carry out the research on "advanced computing technology for AI application." With the furtherance of research work, breakthroughs will be made theoretically and technologically, step by step. Efforts will be made to enable China's AI computer system, around the end of the century, to feature its own distribution of functions and to make avail functions such as computing, sensing, memory, reasoning and learning, the ability to interact with Chinese phonetics, Chinese characters, graphics and images, and plentiful software design and production, so as to make a higher functioning environment.

4. In the High Technology Development Program, the arrangement for the research of laser technology is to carry out frontier research and applied research in a new type of laser science and technology that has outstanding features, high power, high energy and high quality. The focal points are as follows:

For the high power solid state laser the aim was to increase its pulse output energy, improve the quality of the light beam, and make it work at double frequency and treble frequency, so as to make the driver able to meet the research requirements for laser fusion and X-ray laser.

With respect to the short wave chemical laser, the iodine oxide chemical laser with great application potential has been developed. The power output of both pulse and continuous wave is up to the order of several thousand watts. There is notable improvement in terms of the quality of the light beam and the stability of the output light beam. The wave length of the iodine oxide chemical laser is 1.315 micron, and it is the chemical laser with the shortest wave length to date. It is highly efficient and accessible to high power and high energy output. Its wave length is fitted to quartz fiber transmissions, and it may be useful for remote distance processing.

Regarding the X-ray laser, China first obtained the VUV wave band laser in 1989. Soon, the X-ray laser was successfully obtained and high level result had been achieved using electron collision excitation mechanism and three-body composite mechanism. The X-ray laser has the shortest coherent radiation, which has unique application in X-ray micrometry, plasma diagnosis and holography.

China has established laboratories for "short wave chemical laser" "atmospheric optics" and "high power laser and physics." In order to encourage research in new concepts and new propositions, it has set up the Laser High Technology Funds and Laser Youth Funds. Laser technology is an area in which China has a strong basis and has good prospects for further development.

5. In the area of the automation of the manufacturing process, China has a history of more than a decade of organized R&D in robotic technology. Some industries have imported a number of robots, which played a role of enlightenment and demonstration. During the eighth five year plan period, as part of the High Technology Research Program, the theoretical robot prototype, in three different categories and in five types, was prepared. The birth of these robots not only made a breakthrough in the field of special robots in China, but also brought along the development of a number of key technologies as well as the manufacturing of parts and components.

Under the planned economic system, robots were mainly used in hazardous environment operations. In the 1990s, with the establishment of a socialist market economy and the change in the industrial operational mechanism, the use of robots is changing. Intensified market competition made industry consider using robots to remold conventional manual production methods to ensure consistency in quality of products, enhance productivity and increase high-tech value-added products.

The scientists believe that in terms of the development of robotic technology and the manufacturing automation, China should pay close attention to the successful experiences and technologies of other countries and also explore technology according to her own circumstances. China should resolutely abandon importation, assimilation, absorption and copying and instead form her own design capability and utilize her own wisdom to set up her own intellectual property rights. Quality is the key. It would be better to develop more slowly and more stably, but to guarantee the quality of prototype.

The prospect of the development of an industrial robot in China by the end of this century or early next century will depend upon the scale and level of development of China's automobile and electronic industries. If by 2010 the production capacity of the automobile industry in China is to reach the millions and the motorcycle industry the ten millions, they will need to adopt robotic production methods.

6. The energy technology research of the High Technology Research Program covers two fields: advanced reactor technology and coal-fired magneto hydrodynamic technology. Major progress was made in both fields with a number of high level results.

Advanced Reactor Technology. With the efforts of the past decade, breakthroughs have been made in research on key technologies of the fast neutron reactor and its design.

1. In respect to research of fuel and fuel components of the fast neutron reactor, the trial production of high density uranium plutonium compound oxide fuel block and 316 titanium stainless steel cladding components has been carried out. The problems of the key technologies in respect to the production of the fast neutron reactor have been tentatively solved.

2. The engineering design for the fast neutron reactor has been carried out thoroughly.

3. The fast neutron reactor is a reactor type that is able to improve the fuel utilization rate, safety and economy. It is of major importance in the strategic disposition of nuclear energy development through the next century. The current experimental fast neutron reactor is expected to be built soon after the end of this century.

Remarkable progress has been made in the design, research and building of a high temperature gas-cooled breeder reactor of 10 megawatts of thermal power, 2 megawatts of electric power and 700°C reactor outlet helium temperature.

There are broad prospects for a high temperature gas-cooled reactor in the field of thermal energy application. The modular high temperature gas-cooled reactor developed by Qinghua University, which is the most advanced design internationally, is safe, simplified and reliable in structure. The reactor will be built by the end of the century. Its completion will be a landmark for China's lead in the field of high temperature gas-cooled reactor internationally.

The fission-fusion compound reactor project has been undertaken by the Southwest Physics Research Institute and the Hefei Plasma Research Institute of Chinese Academy of Sciences. The compound reactor is an intermediate stage of nuclear fusion, which will be a fundamental way to finally settle the conflict between energy supply and demand.

Coal-fired Magneto Hydrodynamic Power Technology. Coal-fired magneto hydrodynamic power is a new technology that has high efficiency and low pollution. To date, an experimental base for research and production of parts and components of the front cycle and succeeding cycle of coal-fired magneto hydrodynamic has been built. Research on parts and components, such as combustion chamber, power generation channel, special spare heating boiler and recovery equipment, has been carried out. Since 1990, when the experiment equipment was first tested, a large number of tests have been made and a number of results achieved. This includes the power generation test, which, since 1994, has an accumulated power generation of 3 hours and 30 minutes, with peak value generation as high as 119.7 kilowatts. This exceeds the design index of power to be generated by the first test power channel. In August 1995, the initial design of the 12 megawatt coal-fired magneto hydrodynamic-steam co-cycle experiment equipment was completed. During the ninth five year plan period, the existing equipment will be used to carry out tests and research the upper stream parts and components of coal-fired magneto hydrodynamic power. This shall bring the research of key technologies up to new levels.

7. Material science is one of the fields with the fastest development in recent years. As part of the High Technology Research Program, a number of important results have been achieved in the past decade in the development of new materials.

A number of important fields in new material research have been opened up. Some examples are photoelectric materials and metal organic chemical compounds (MO), preparation and application of diamond thin film, intermetallic compound materials, new technology of modified materials and surface optimization.

China's superiority in artificial crystal has been consolidated and developed, which keeps the country in a lead position in the field of inorganic non-metal non-linear photocrystal and has closed the gap with respect to laser crystal.

A number of major research results have been commercialized, such as the nickel-hydrogen cell and the high performance, low temperature sintering multilayer ceramic capacitor (MLC), which has furthered the commercialization and industrialization of the High Technology Development Program.

Defense construction has been provided with key new materials, such as infrared camouflage material and new solid state propellants.

The technological level of equipment has been improved by relying on China's own technical force, as in the development of the automatic YAG single crystal growth furnace, direct current plasma spraying diamond thin film growing equipment and the omnibearing ion injection equipment.

The implementation of the High Technology Research Program in the past decade has proved that it is of major importance to begin high technology research in China. In the process of implementing the Program, quite a few characteristics have surfaced, as discussed hereafter.

The development of high technology in China had been focused on limited targets, but the selection of targets was under careful consideration. The selection was aimed at the world frontier of high technology development and considered the future prospect of development. At the same time, the selection took into consideration the country's actual circumstances and considered economic prospects. So far as careful selection of targets, the implementation was in good form and the results were remarkable.

As the Program has taken it as an important goal for further development in the next century, it is of major importance to train the scientific talents of the new generation. Therefore, all projects of the Program included a number of middle-aged and young scientists as playing a crucial role.

The High Technology Research Program is a State program and is supported by State appropriation. However, in order to ensure efficiency, the operation and management of the Program was quite different from the original style of research management under the planned economy. Under the planned economy, research projects were assigned to different departments according to the fields of research, and State appropriation was distributed similarly. The High Technology Research Program adopted a different way. The distinction between different departments was broken up. Universities and research institutes under the charge of different departments would apply to implement projects on competitive basis. The assignment of a project was on a selective basis. An expert committee played a major role in decision making. All these measures were aimed at optimizing the utilization of the best scientists, best institutions, and limited State appropriation, to achieve the highest efficiency.

By the end of 1994, right before the 10-year High Technology Development Program ended, the State Science and Technology Commission and the Chinese Academy of Sciences jointly sponsored a conference to address some major issues: To identify China's high technology development strategy for the early twenty-first century; to put China's high technology research and high-tech industry in the world's forefront and to take a position in specific fields. Song Jian, chairman of the State Science and Technology Commission, made an important statement at the conference.[1] According to Song, it was necessary to re-evaluate and readjust the strategic goal of the 863 Program. When China formulated the High Technology Development Program in 1986, it was during the Cold War era. After 1992, the State policy was readjusted. Song said, "The principle of taking economic construction as the primary concern was decided. Our strategic goal is to improve economic strength, productivity and the people's living standard of our country. Deng Xiaoping introduced the thesis that science and technology be the first force of production in 1988 and again, in 1991, indicated that it is necessary to develop high technology and realize industrialization. This provides a basis for readjustment of the strategic goal of the 863 Program." Song pointed out: "If we want to improve productivity and people's living standard, we should improve our economic strength and make efforts to grasp the development of high-tech industry." It was under this premise that Song stressed the issue of innovation. He said: "In the past, we kept saying that the 863 Program was to follow the frontier of high technology in the world. There was a particular reason to use the word 'follow'. As a policy, it is better to change it into innovation now. It is necessary to establish a technology innovation mechanism, and innovation requirement should penetrate throughout the whole process of R&D. Without innovation, we won't have our own high-tech industry. We can't develop a high-tech industry without innovation."

It was under this guiding thought that the second term of the High Technology Development Program was formulated. Some experts preferred to described it as the "Super 863 Program." By all means, high technology research has become an important strategic issue in China, and developing a high-tech industry has become a more urgent issue in China's S&T policy as well as in China's industrial policy.

NOTE

1. Song Jian, former State Councillor and Chairman of the State Science and Technology Commission. Song's statement was addressed to the 28th Xiangshan Scientific Conference, co-sponsored by the State Science and Technology Commission and the Chinese Academy of Sciences. The Conference was focused on the development strategy of China's high technology in the twenty-first century. *Science and Technology Daily* had a report on the conference on January 4, 1995.

17

The Torch Program and the Progress of High-Tech Industry

The High Technology Development Program is a research program with mid- and long-range goals. Soon after the Program was organized in 1986, the Torch Program began in 1988. Though both are high technology development programs, there is a big difference between the two. The Torch Program is aimed at the commercialization, industrialization and internationalization of high and new technology. Therefore, it is market oriented. The projects of the Torch Program are not necessarily on the frontier of the development of high technology. But it must be feasible for its products to grow into marketable commodities, to develop in volume production and to form high-tech industries. Some of the Torch Program projects may not exactly be in the high-tech category, but if the technology is new to China, it may be included in the Program. Thus, the term "high and new technology" originated.

In the late 1980s, China possessed both the need and the capability of starting her high technology industry. In the coastline areas, under the coastline area economic development strategy, export-oriented industry had rapid development. When striving to upgrade industrial levels and gradually replace labor intensive products with those that utilize high-tech, it was necessary to actively develop new technology products and a high technology industry, to direct the industrial formation to a higher level without missing opportunities. After several decades of economic development, China had built an initial industrial system with various industrial departments and had a considerable base in the fields of space technology, nuclear industry, computers, biotechnology, lasers, electro-mechanical industry and new materials with the capacity of a whole system of research, experiment and production as well as a relatively strong product development capability. During the late 1980s, China had a scientific and technical force of 8 million. In addition, with the establishment of special economic zones, the coastline cities had many economic and trade linkages with foreign countries, and the scope of economic and business exchange and cooperation with countries abroad had been expanding constantly. Conditions were available to start the development of a high technology industry. Therefore, as early as the end of 1986, the government indicated in the "Outline of High Technology Development Program" that "it is necessary to follow the world advanced level in a few select important high technology fields and to set up necessary high technology industry." In October, 1987, it was put forward in the Thirteenth National Congress of CPC that "In order to reasonably readjust and remold industrial formation, we should take the remolding and developing of conventional industry with advanced technology as focus and, at the same time, pay close attention to developing rising high technology industry."

The coastline areas and central cities were the best places for the development of the high and new technology industry. The development of a high and new technology industry had been seen as an important part of the coastline economic development strategy. The development of a high technology industry should rely on the existing industrial base and scientific and technical force. Large and medium-sized industries had strong economic and technical strength. However, the large and medium-sized industries, aside from their heavy responsibilities in fulfilling existing production tasks, were still unable to free themselves from the long standing influence of a central planning economic system, while the establishment of a new flexible and invigorated operational mechanism through reform was still in its initial stage. In this regard, new science-technology-oriented enterprises had superiority. Though these enterprises were small in production capability, they were strong in their technical force and flexible in their operational mechanism. The township industries also had considerable economic strength. Therefore, development should make good use of various kinds of enterprises and enable them to maximize their superiority in developing high-tech industry.

On the other hand, the scientific and technological community was an important force to rely upon in developing high and new technology industry. There were more than 1,000 research institutes under various central ministries with 300,000 scientific and technical personnel. This was in addition to the research institutions located in various provinces and municipalities that had a strong technical force. In the meantime, a huge potential for a scientific and technical force resided in higher education institutions. In China, more than 800 universities and colleges were devoted to science, engineering, agricultural and medical education. Should the necessary conditions be available and encouragement be given to research institutions and higher education institutions to develop high and new technology products, to start up science-technology-oriented enterprises or to combine with various kinds of industries, these institutions would be a new vital force not to be neglected in the development of high and new technology industry.

In view of this, the Torch Program put forward a few guidelines for developing high and new technology industry:

1. To develop high-tech products and start up high-tech science and technology enterprises to support domestic and international markets by relying on the scientific and technological community and maximizing its superiority, in order to promote the economic development of China's coastline areas. At the same time, it is necessary to organize the scientific and technological community to provide large and medium-sized enterprises or township industries with the necessary technical support in developing high-tech products.

2. In the process of implementing the Torch Program, to organically integrate the development of high and new technology industry with furthering the reform of the science and technology system; to promote the integration of science and technology with economic development in coastline areas by further relaxing control over the management of research institutions and scientific and technical personnel, so as to open up new paths for science and technology growing into the economy.

3. To pave the way for commercial use of the results of the High Technology Research Program and other research programs, including new inventions and patented technologies, to help speed the pace of the productive transformation of various kinds of valuable scientific and technological results.

4. To provide direction and assistance to scientific and technical personnel who intend to devote their efforts to starting up high technology industries and to bring up thousands of good entrepreneurs with technical expertise, as well as management and operation capabilities, and other kinds of senior experts and gear them to the needs of developing export-oriented economy. In particular, bring up a number of successors from the younger generation to engage in the development of the high technology industry.

In the light of these guidelines, a series of policies for the development of the Torch Program were formulated. The major contents were: To be oriented to domestic and international markets, to take products as "dragon head" (as a locomotive for a train), high and new technology as support and the formation of a scale economy as the goal. This would serve to constantly enhance the competitiveness of high-tech products, enable the high-tech industry to make up a larger proportion of the whole industrial formation, make the output value of high-tech industry possess a larger proportion in the total output value of the national economy and bring export products of high-tech industries up to a level that would match that of the medium-level developed countries in the early 1980s.

Since it was not the plan to develop high technology industry extensively, it was important to select the fields for development. The Torch Program had taken microelectronics, computer, information and telecommunications, biotechnology, new materials, electromechanical integration, lasers, new energy and highly efficient energy saving as the primary direction for development.

Research institutions and universities were encouraged to start their own science-technology-oriented enterprises and, at the same time, form high and new technology industrial combined entities with industry at different levels and in various forms.

The Torch Program proposed the opening of various channels to raise capital, including direct seed money from the government, loans from banking institutions and issuing stocks and bonds. The banks were required to set up special science and technology loans. Areas with the proper conditions might set up science and technology investment institutions. Attracting foreign investment was also encouraged.

Flexibility would be kept in ownership of start-up high and new technology enterprises. Such enterprises might exercise various kinds of ownership.

As the high-tech industry is knowledge intensive with a high ratio of scientists and engineers to workers and as the fast renewal of high-tech products and the rapidly changing market requires highly sensitive reactions from the high-tech industry, a flexible operational mechanism to adapt to changing markets and a good information network, there should be good infrastructure and an environment ready to facilitate starting and developing a high-tech industry. It was in this regard that the Torch Program put forward a series of measures, including (1) encouraging research institutes and universities to start their high-tech science and technology enterprises, supporting the combining of research institutes and universities with industries, as well as facilitating the small and township industries to develop into high-tech industries and (2) encouraging the establishment of technology development zones and high-tech business incubation centers. The former would be in favor of integrating industry with research and accelerating the development of high-tech industries, as well as promoting local industrial reformation and raising economic prosperity of the localities. The latter would offer comprehensive services for start-up high-tech enterprises, services that would be badly needed by the research institutes, universities and individuals when starting up their own high-tech businesses, especially when scaling their operation up or introducing their products to the market.

The Torch Program, instead of being a scientific research program, is actually an umbrella that directs the development of high and new technology industry in China and attracts a large number of scientific and technical personnel to the development and industrialization of high and new technology. What the Torch Program offered is mainly a series of guidelines, policies, overall planning, review and approval of projects, which is followed up with policy support. In the process of implementing the Torch Program, the initiatives of research institutions, universities and various localities have played an important role.

As the Torch Program encourages research institutions and universities to start their own high-tech businesses, a number of science and technology enterprises have started with the backing of research institutions or universities. In the past decade, quite a few important high-tech businesses that started with the backing of research institutes have grown to be large high-tech enterprises. The Legend Group is a good example.

Legend is a famous computer company in China. It started with backing from the Computer Technology Research Institute of the Chinese Academy of Sciences. When it first started in 1986, it received starting capital of 200,000 yuan from the Institute. The company had a staff of 11 members and possessed two office rooms. Now, in eleven years, it has grown into a company with total assets of U.S. $600 million. Since 1996, Legend has ranked first in China's domestic PC market. In 1997, its domestic market share came to 12% and achieved total sales of 12.5 billion yuan. In 1998, Legend is striving to attain annual sales of 18 billion yuan. Legend is planning to achieve annual sales of U.S. $3 billion by 2000 and become one of the world's 50 largest computer companies.

Universities are an important force in developing a high technology industry. Since the early 1990s, some of the key universities have established various kinds of science parks or science and technology industrial parks, such as the Shanghai Polytechnic University Science and Technology Park, the Northeast University Science Park, the Harbin Polytechnic University High and New Technology Park, the Qinghua University Science and Technology Park, the Nanjing University Science and Technology Industrial Park, the Xian Jiaotong University High and New Technology New Venture Center, the Southeast University High Technology Industrial Group and the International Science Park jointly sponsored by Wuhan University and the Hong Kong Coastline International Group. The objective of establishing science and technology parks is to use these parks as incubators or bases for trial production, to speed up the industrialization of major scientific and technological results, to promote technology renovation in specific industrial fields and to speed up the renewal of products by way of "incubation" or "radiation" in order to contribute to the economic construction of the country.

As one measure of the scientific and technological reformation as well as educational reformation in China, universities are encouraged to start their own science-and-technology oriented enterprises, by either starting on their own or starting jointly with industrial enterprises. The science and technology enterprises started by universities make up an important part of non-governmental science and technology enterprises, and played an important role in the growth of high and new technology industry. By 1994, the enterprises started by universities came to 1,797, with a full time technical staff of 34,046, including 2,998 senior scientists and engineers from universities. The universities, in total, invested 620 million yuan and had achieved total sales of 7.3 billion yuan, with more than 900 million yuan in after-tax profit. All these enterprises had accepted more than 350,000 student interns. A considerable part of these enterprises are, for the time being, located in different high and new technology industrial development zones.

More enterprises were started by polytechnic universities or engineering schools. Some enterprises of this category have become well-known both domestically and internationally, such as the Ziguang Group of Qinghua University, the Chemical Engineering Separation Technology and New Stuffing Development Research Center and the stuffing factory of Tianjin University, Shenzhen Tianji Photoelectron Technology Industrial Co. Ltd. of the Electronic Science and Technology University. The comprehensive universities are science oriented, with more basic science courses and more basic research, as well as high technology research. Though comprehensive universities started fewer enterprises than polytechnic universities, many of them have high technology enterprises with large sales and a higher after-tax profit, such as the Fangzheng Group of Beijing University, Shanghai Fuhua Industrial Co. Ltd. of Fudan University. In the public appraisal, "China's One Hundred Giants of High Technology Enterprises," the Fangzheng Group of Beijing University, Fuhua Industrial Co. Lt. of Fudan University and Ziguang Group of Qinghua University ranking 21, 62 and 83 respectively.

The high technology enterprises started by universities are stronger in the sense of innovation. This has been a weak point of many industries in China. What prevails in some industries is copying foreign products. A number of technology results were virtually copied from products sold in large volume on foreign markets. While China spends a lot of time turning out a prototype, foreign manufacturers now sell their new generation products on the market. Therefore, even by making a substantial investment in transforming the prototype into commodities, it is impossible to compete with new foreign products on the international market. In this regard, the Fangzheng Group is an extraordinary example. Innovation has been their strong point. In the process of developing photocomposition technology, the Fangzheng Group insisted on innovation and a high starting point. Two decades ago, the Group saw a new opportunity. During that time, what prevailed in Japan was the photomechanical type of second generation photocomposition machine; and what prevailed in Europe and the United States was the cathode-ray tube third generation photocomposition machine. As early as 1976, the predecessor of the Fangzheng Group made the decision to skip over the second and third generations and directly develop a fourth generation laser photocomposition system. The laser photocomposition system didn't prevail internationally until 1985. Considering the large number of Chinese characters and the large volume of information of character pattern, the Fangzheng Group adopted a unique method that solved the problem derived from a large volume of information: the mathematical method of information compression. It was successful as soon as it appeared in the marketplace. It took an overwhelming majority of market share in terms of Chinese character photocomposition systems both domestically and internationally.

In 1991, the Fanhzheng Group seized another important opportunity: the color publishing system market. In the past twenty years, China's color publishing field had been monopolized by products of four electronic color scanners manufactured abroad. China spent 20 years trying to copy the foreign color scanner; but although it succeeded in copying one generation of color scanner, it soon failed because of the newer generation of products from abroad and was unable to get into the market at all. The conventional electronic color scanner was a specialized closed system with a high cost, while the color laser photo composition system of the new generation was far better in terms of cost, flexibility and functions. In particular, it was better in the processing function of integrating language and graphics. Of course, all systems from abroad were unable to process Chinese characters. Seizing this opportunity, in January, 1992, the Fangzheng Group let its color laser photo composition system be used at Macao Daily to replace the electronic color scanner. Using Fangzheng's new system, Macao Daily was able to publish 4–6 color pages each day. This was the first time the world could realize the process of the color picture integrated with the Chinese language as well as put a whole color page on newspaper. In the past, using the manual clip and paste method to produce a color page, it took an hour and a half to two hours to finish each page. Now, it took only a little more than 10 minutes to output a whole page with four color separation pictures, so that it was now possible to publish color pictures on news pages and meet the high requirement for timely reports. Later, the Fangzheng

Group improved the quality of the Fangzheng plate making color publishing system, so it was able to match that of the electronic color scanner and completely replace it. Then, the Fangzheng Group once again, through innovative design, developed a special large scale integrated circuit of 0.7 micron line width to speed up the process of color pictures, graphics and characters to a level that made it the world leader. In the past, the Fangzheng Group relied on its superiority in processing Chinese characters to meet the demands of domestic and international markets. Now, the Group is able to compete with products abroad in the processing of color images and graphics and in the publishing field.

The experience of the Fangzheng Group proved that it is beneficial to establish high-tech industrial enterprises with universities. Beginning with innovative research, then transforming to product and finally giving birth to a new high-tech industry—that's the typical process of a successful high-tech industrial enterprise. Within the Fangzheng Group of Beijing University, the entire process is available for covering all links of research and development. The upper stream of the process is covered by the State Key laboratory of Linguistic Information Process Technology, the middle stream is covered by the State Electronics Publishing New Technology Engineering Research Center and the lower stream is covered by the Fangzheng Group Co. Either the lab and engineering center or the company are located on the same campus. The Fangzheng Group is the integration of scientists with a business mentality and entrepreneurs with a scientific perspective.

The integration of industry with research institutions and universities was encouraged by the Torch Program with a view to bringing a research force together with the superior productive conditions in the industry. The policy promotes combination of a number of universities and research institutes with industries and has brought about successes in the industrialization and commercialization of innovative products. The success of MDI, jointly developed by Qingdao Chemical Engineering College and Wanhua Corp. Group in Yantai, is an outstanding example.

MDI, as a high technology, is mainly applied in producing hard bubble used in heat insulation and heat preservation for refrigerators, cool storage, oilfields, central heating and so on. It also involves manufacturing highly elastic resilience material, the half hard bubble and adhesives. The applications are distributed among electromechanical, ship-building, aeronautics, automobile, construction and textile fields.

MDI was monopolized by corporations in the United States, the United Kingdom, Germany and Japan. Four industries in China were approved to develop MDI technology through importation. The four industries tried to import a whole set of MDI production equipment with an annual capacity of 20,000 tons, but all proposals for technology acquisition were turned down by foreign corporations that owned the technology. The only way around this was to develop it using China's own technical force. In 1996, Wanhua Corp. Group and Qingdao Chemical Engineering College, through quite a few years of joint research and development, successfully developed a whole set of MDI production equipment with an annual capacity of 20,000 tons. The cooperation started in 1988, and in 1989 improvements were carried out on the imported equipment of a key technology of MDI—the phosgene process—and brought the MDI phosgene load to 105% by that summer. This was unprecedented. The improvements made it possible to carry out stable operation of the MDI production equipment and achieved a 10% increase in annual output and increased the total annual output to 9,000 tons. In 1993, the Wanhua Corp. Grop again cooperated with Qingdao Chemical Engineering College to develop MDI technology. During the period of 1993–1995, the annual output of MDI came up to 10,000 tons. In April, 1996, through technological improvements, the MDI production line operated at an annual output of 13,000 tons. At the same time, both parties had successfully developed a new set of MDI production equipment for an annual output of 20,000. In addition, the core technology—phosgene technology—for the development of MDI production equipment that had an annual capacity of 40,000–80,000 tons had been mastered.

The achievement made by the successful cooperation was due to maximizing the superiority of both the university and the industry. The industry was the backbone of the joint development. The project was funded by the industry. Total investment for the development of MDI technology came to 15 million yuan, including 3.65 million yuan directly allocated to the Qingdao Chemical Engineering College. The investment was made by the Wanhua Corp. Group. The idea of the technology development was put forward by Wanhua Corp. Group, as it came from the requirements of practical production, and the results were examined by the process of production in industry and eventually applied to industry. It is impossible for technology development for scale production to come from the lab of a university. It must come from industry. However, this doesn't mean that the university plays only a minor role. The reason the Wanhua Corp. Group chose Qingdao Chemical Engineering College as its partner was that the College had developed a computer engineering chemical simulation system (EEES), which not only took the lead in the country and won the highest award from the Ministry of Chemical Industry but had also been applied by more than 50 manufacturers in petroleum, chemical, pharmaceutical and light industry.

Large enterprise is an important industrial force not to be neglected. Though many State-owned large enterprises suffered from serious deficits in the past few years, many of these industries had a stronger industrial base and technical force that favored developing high-tech products.

Some large industries did take advantage of their strong position and achieved good results in high-tech product development. Changhong TV is a good example. The success of Changhong is based on its strong technical force and successful operation, which maximize its technical superiority. Changhong is taking the lead in China's television industry with a dominant share of the television market. Changhong Television Manufacturing Group is a State-owned large scale enterprise that has grown out from the Changhong Machinery Factory (a State-owned large enterprise established in the 1950s as one of the 156 key projects aided by the Soviet Union). The Changhong Machinery Factory had been an important base for the development and production of the airborne fire control radar system and had strong technological development capabilities and technical base. At the end of the 1970s; Changhong took the lead in implementing the development strategy of integrating military production and civil production. It selected the television product line that had broad market prospects and, at the same time, was able to maximize the technical superiority of military production. When Changhong stepped into the field of television manufacturing, it made its technology, technique and experience in quality control (which it had accumulated in the process of producing military products) available for the development of television. In the mid- to late 1980s, it had twice cooperated with National of Japan, importing an advanced production line with manufacturing technology as well as tubes to start volume production of the color TV. Since first turning out its TV products, Changhong has insisted on the strategy of "while carrying on production of the current generation, go ahead with a R&D of the next generation, prepare the reserves for the generation after, and start pre-study of future generation." In technology development, Changhong covered three stages: importation, joint development with a foreign partner and its own development. Hence, Changhong formed its own comprehensive development capabilities.

Since the 1980s, in the production of the TV set, Changhong has developed its products from a single type of core unit and single variety to a series of products with multiple type of core units and multiple varieties. Now, it has transformed from importing whole sets of core unit technology into developing it through its own efforts with the advanced chips it imported internationally. Changhong has successively developed a dozen types of core units. Quite a few of them are used for large screen TVs designed with various advanced technologies that have emerged internationally. This makes the quality of image, audio, safety, reliability and life span superior to the State standard. In addition, as early as the 1980s, Changhong continued R&D on digital technology and, since 1990, has applied it to its products. Now, Changhong has adopted various digital circuits, such as the bus control circuit, digital comb filter circuit, NICAM decoding circuit, PIP circuit, digital Karaoke circuit and the all digital image quality processing circuit. Changhong has also mastered the core technology of color TV—the software technology—which makes Changhong able to expand its performance in satisfying various demands of customers. This also displays the development capability of Changhong on its own accord. The capability of developing powerful technology enables Changhong to increase its competitiveness.

The success of Changhong is not only due to its strong industrial and technological basis, but also to its efforts to further its technological superiority. Under Changhong's R&D system, there are a State key technology center, four product development design institutes and a production technology research institute. Aside from updated research and experimental facilities, there are more than 2,000 engineers and technicians directly involved in product development. Changhong has developed close cooperative relations with quite a few famous universities and research institutes in the country, including Qinghua University, the China Electronic Science and Technology University, the Sichuan United University, Chongqing University, the Research Institute of Electronic Industry and the Broadcasting Research Academy. In recent years, Changhong has increased its technological development investment by a big margin. In 1996, its technology development expenses totaled 420 million yuan, which accounted for 3.89% of annual sales. In 1997, it increased to 800 million yuan, which accounted for 4% of annual sales. With further development, Changhong is expected to increase its technology development expenses to over 6% of its annual sales by the end of the century.

One of the notable features in high-tech industrial development is that a number of small or township industries have entered into the high-tech field, and some of them have been quite successful in growing into high-tech enterprises. Quite a few township industries have grown into large high-tech manufacturers and take the lead in specific fields of high-tech products. The experience of the Glanz Group is a typical example.

Glanz is now one of the giants in China in manufacturing electrical home appliances. It was developed from a small township industry. The small township industry was first established in 1978 to produce eiderdown outerwear. During that time, it had only 200 workers engaged in washing goose and duck feathers by hand to supply the export company and had annual sales of one million yuan. In 1983, the small township industry entered into a joint venture with a Hong Kong firm and the Guangdong Animal By-Products Import and Export Company. They established a wool manufacturing industry equipped with a brand new type of roving production line, which was imported from Japan, and achieved annual sales of more than 30 million yuan. Later, it successively cooperated with a Hong Kong firm and a U.S. firm to produce down robes and down garments for export. In 1988, it put all its joint venture businesses together and formed the Guizhou Animal By-Products Group, which achieved annual sales over a hundred million yuan. In 1992, Guizhou Animal By-Products Group was renamed the Glanz Group.

As early as 1991, the chief executive of the company had spent a year conducting market surveys and found that the domestic microwave market was dominated by imported goods and that there would be attractive prospects for the microwave market in the country. The Glanz Group made the decision to get into the microwave market and spent only half a year completing all procedures, including project registration, and importing equipment for production. Through a technology cooperation with Toshiba and importing an updated automatic production line, the China-foreign joint venture of the Glanz Group was established and started its trial production in September, 1992. Right from the beginning, Glanz focused on production and marketing of electrical home appliances, including microwaves, air cleaners and electric pots. In 1993, Glanz started to carry out an industrial reformation and transformed from a textile manufacturer to a electrical home appliance manufacturing group. In 1993, the Group turned out 10,000 microwaves in its trial production and took a leap forward as a manufacturing group with diversified products. The Group, with nine subsidiaries, had transformed itself from a traditional labor intensive industry for the production of down robes and down garments into a technology intensive industry. Glanz had constantly developed its market and formed a transregional, trans-industrial-departmental diversified manufacturing group.

The year 1995 was important for the growth of Glanz. A breakthrough made in that year allowed Glanz to double its annual output of microwaves and achieve an annual output of 200,000 microwaves and increased its market share to 25.1% of the country's total microwave market. It ranked number one in the microwave industry in the country. In 1996, more intense competition took place in the microwave market. In order to ensure further development, Glanz decided to cut the price of the microwave by a big margin and expanded its market share to 34.5%. Once again it ranked first in the microwave industry in the country. Its annual output in 1996 came to 6,550,000 sets and formed a microwave production base. Aside from developing its domestic market, Glanz is making efforts to develop its international market. Glanz microwave is sold in about 100 countries and obtained ISO 9001 certification and the UL certificate of the United States, the GS certificate of Germany and the CCE certificate of Denmark, Norway and the European Community. This helped to lay a solid foundation for Glanz to enter the international market.

The experience of the growth of the Glanz Group is typical of a number of high-tech industries that grew from township industries. Glanz is located in Shunde, Guangdong Province, which is home to quite a few electrical home appliance manufacturing giants. Quite a number of township industries, after accumulating their financial power through labor intensive and low technology content manufacturing, either absorbed technical power from research institutions and universities or imported advanced foreign technology and developed into high-tech industries. So far as these industries are flexible in organizing their product line according to changing market demands, they have been successful in achieving considerable market share in their respective industries and have grown into giants in quite a few high-tech product fields.

The science and technology enterprises started by universities and research institutions have played an important part in the growth of high and new technology industry in China. In the 1990s, a number of these industries have grown either in scale or in scope of operation. In the meantime, especially after 1992, when China decided to further the opening policy and broaden the scope of opening up to the outside world, a number of global corporations have come to China to develop joint ventures or to establish their production bases in China, many of them in the technology development zones. In the 1990s, for a period of time, there was overcrowding in developing technology development zones. Many localities established their own development zones, but quite a number failed to attract any science and technology enterprises. With only enthusiasm to pursue high technology and desire to enjoy the preferential treatment given to technology development zones but without the necessary conditions, the efforts led nowhere. Later on, after a readjustment and rectification had been carried out, a number of zones without the proper conditions were canceled. The readjustment put the remaining development zones on a more solid base. In quite a few zones, domestic high technology enterprises are now developing side by side with the transnational corporations. For instance, in Shangdi Information Industry Base in Beijing, the domestic giants such as the Fangzheng Group and the Legend Group are in the same neighborhood as IBM and Mitsubishi.

With the development of high and new technology industry, the technology development zones are playing an important role in improving the regional industrial structure and promoting regional economic development. For instance, the Tianjin Economic Technology Development Area is one of the major development zones in the country. With eleven years' development, up to 1995, it had accumulated 2,557 approved enterprises with foreign investment, and the total agreement investment came up to U.S. $5.869 billion; while the total number of domestic enterprises was 8,442, the total registered capital amounted to 16.264 billion yuan. An industrial system has been formed in the development zone in the fields of electronics, machinery, food, bio-medicine, fine chemicals and building materials. The Development Zone has constantly made efforts to attract projects with high technology contents and high value-added features, so as to promote industrial upgrading and technology innovation. As more transnational corporations have come to invest in the Development Zone, the level and quality of the projects have improved rapidly. Among the Fortune 100 corporations (1993), 15 have invested in the Development Zone. Quite a few world-famous transnational corporations, such as Motorola of the United States, SEW of Germany, Samsung of South Korea and NOVODISK of Denmark, have each invested over a hundred million U.S. dollars in the Development Zone and set up their Asian manufacturing centers. By the end of 1995, there were 23 enterprises in the Development Zone that had been identified by the government as high and new technology enterprises, including NOVONORDISK, which produces industrial enzyme preparations and insulin; American Standard, which produces ceramics; MEITE, which produces modern packaging products; and AST Computer and Motorola, which have integrated circuit products as their main line.

Tianjin, as one of the major large cities in China, has a long history; as many of its industries are outdated, the economic growth of the city has slowed down. In contrast to the conventional industry of the city, the Tianjin Economic Technology Development Area is playing a more important part in the economic growth of the city. Currently, its total industrial output value accounts for 16.5% of the whole city and the newly increased industrial output value accounts for 50% of that of the city, which makes the development Zone the largest economic growth point of the city. According to estimates, by the end of the century the total industrial output value of the Development Zone will account for 35% of that of the whole city. The accelerated development of the Development Zone will play a decisive role in rejuvenating the old industrial base of Tianjin.

The situation in Tianjin is typical of a number of major technology development zones in the country, such as Shenzhen, Zhuhai and Xiamen. The rapid growth of coastline cities and development zones has largely depended on the export of labor intensive products, but now the market is saturated, and labor costs have been enhanced significantly. Now, there is a tendency to transfer labor intensive industries to inland areas. For instance, in Shenzhen, more than 600 labor intensive industries have been transferred to inland areas. Instead, Shenzhen is trying in every way to develop high and new technology industry. Insofar as Shenzhen has the advantage of access to international markets but doesn't have a strong knowledge base, it is developing various combinations with inland cities and setting up joint scientific and technological development projects to transform technological results into high technology products and enter into international market. For the brand-name high and new technology products in Shenzhen, the manufacturers have made full use of their inland supply sources for parts and components. They have arranged for a number of specialized factories to engage in producing parts and components in large volume and with high efficiency to fit the mainframes made in Shenzhen. What has begun in the development zones will be the general tendency of the whole coastline area—to introduce more high and new technology industry and replace labor intensive industry. Labor intensive industry will be transferred to inland areas. For the time being, more high and new technology industries are being introduced into the coastline areas and central cities in China through joint ventures with foreign businesses, or as a production bases of transnational corporations, than those developed by China's own innovation. In the coming years, the scope of introducing high and new technology industry will be continuously expanded, and domestic innovation will grow correspondingly.

18

Non-governmental Science and Technology Enterprises

Non-governmental science and technology enterprise is a compound name to cover a group of enterprises in different forms and with different ownership but with a common nature: oriented toward science and technology. Sometimes they are also called science-and-technology-oriented enterprise.

All science and technology enterprises focus on technology research, development and service. The ownership of these enterprises is diverse: Some are spun-off from and backed by their parent institutions, some are run by a cooperative group, some are run by individuals and some are private. However, these enterprises don't receive government appropriations. The group of people who get together on a voluntary basis raise capital by themselves, operate on their own, are responsible for profits and bear the risks. In a country like China, which had been used to a planned economic system, the emergence of a group of non-governmental science and technology enterprises was something brand new. However, the new enterprises soon gained momentum in the market. As a newborn thing, the non-governmental science and technology enterprise, when it first appeared in the 1980s, broke the bondage of the old system and opened new paths in China.

The new non-governmental science and technology enterprises saw rapid growth. As early as the beginning of the 1980s, under the influence of the reform policy and opening up to the outside world, a small number of scientific and technical personnel had started the first batch of non-governmental science and technology enterprises. By the end of 1984 and early in 1985, after decisions on economic reform and scientific and technological reform were issued by the central government and it was officially announced that it was permissible for the collectives or individuals to set up technology research and technology service institutions, the development of non-governmental science and technology institutions accelerated. In early 1986, the State Council issued "Some Stipulations on the Further Promotion of Reform of Scientific and Technological System," in which it was announced that certain measures would be taken to further relax the management of scientific and technological personnel and provide them with more flexibility. In the meantime, the State Science and Technology Commission approved the creation of the China Non-governmental Science and Technology Entrepreneurs Association. Immediately after that, local associations were set up in 18 provinces, municipalities and autonomous regions. Since then, the development of non-governmental science and technology enterprises has reached a new stage. According to an investigation in 10 provinces and municipalities in March, 1987, there were 2,304 non-governmental science and technology enterprises in these provinces and municipalities, comprised of 53,476 employees; and in June, 1988, in the same provinces and municipalities, the non-governmental science and technology enterprises had increased to 3,760. According to another investigation, conducted at the end of November, 1988, there were 11,196 non-governmental science and technology enterprises all over the country.

The significant growth of non-governmental science and technology enterprises was due to the development of a new operational mechanism and management system, which brought about more effective integration of science and technology with the economy.

The decision-making process is the most important link in the operational mechanism of an enterprise. Unlike State-owned enterprises, non-governmental enterprises didn't need to wait for a senior authority to make decisions. The decision-making power was retained by the internal management team of the enterprise. According to a sample investigation of 22 non-governmental science and technology enterprises in Beijing, those which made decisions on their own, right from the selection of research projects through the sourcing of materials, had risen to 99%. Either to gather market information and feedback or to develop technology and products, fast reactions are required in order to operate a modernized science and technology enterprise. Only by making decisions on one's own, as the non-governmental science and technology enterprises did, could one meet the requirement of fast reaction.

In order to survive and develop, the non-governmental science and technology enterprises first of all selected their R&D projects carefully and meticulously, so as to form their business operation on a solid base of science and technology. They looked to science and technology as their guide to promote production and carry out technology products trade. The integration of technology and industrial manufacturing with trade, with technology as the main, was an operational mechanism that was able to shorten the R&D cycle, to overcome the inability to transform technology results to commodities, to help materialize science and technology and to transform it to realistic productive force quickly.

Slow reaction to changing market demands had been a serious weakness of State-owned enterprises. They were used to a planned economy and found it very difficult to adapt themselves to a market economy. Insofar as non-governmental science and technology enterprises followed a flexible mechanism, they were able to react to the market demands sensitively and quickly and to operate under the mechanism of market-information (feedback)-development-market, thus forming a sound cycle with the market as the guide. According to an investigation of 20 science and technology enterprises, the cycle from R&D to commodities entering into markets was no more than one year in 75% of projects undertaken by these enterprises.

The presence of talented people is a most important factor for the success of a science and technology enterprise. The criteria for talented people in non-governmental science and technology enterprises were an ability to learn and possession of operational talent. Personnel management followed the principle of two-way selection and selecting excellence. The personnel of these enterprises kept rotating in order to optimize the composition of the technical staff.

The distribution of non-governmental science and technology enterprises was linked to economic benefits. China's State-owned enterprises were used to an egalitarian distribution system that did not include incentives to workers. One of the major contents of China's economic reform is to free the distribution system from the egalitarian principle, but this was very difficult to realize, due to force of habit. The non-governmental science and technology enterprises had broken the force of habit and exercised a new distribution system that closely linked the income of workers and staff with the profits of the enterprise and placed the income of individual worker or staff member under a more work, more pay principle. Hence, both competition and the incentive mechanism were introduced into the distribution system of non-governmental science and technology enterprises.

As a good economic behavior, the non-governmental science and technology enterprises all created their own science and technology development funds, social insurance funds and welfare funds. Attention was paid to accumulating funds, the increase and renewal of fixed assets, technology renovation and investment efficiency. The reserved profits were mainly used to develop projects with good prospects, which prevented them from being spent on premature consumption and short-term behavior. All this boiled down to one most important point: non-governmental science and technology enterprises were able to create high economic benefits by relying on technology progress. Self-accumulation, self-development and self-restriction mechanisms had been established in the non-governmental science and technology enterprises.

Scrutiny of the growth of Zhongguancun Electronic Street may help us have a clear understanding of how non-governmental science and technology[1] enterprise emerged, how it developed and its characteristics and significance for China's economic and scientific and technological reform.

The Zhongguancun Electronic Street in Beijing may be the most well-known place of a concentration of non-governmental science and technology enterprises. As early as 1980, a research fellow working with the Institute of Physics under the Chinese Academy of Sciences came out from the research campus to start up the first non-governmental business on the street. It was named "Advanced Technology Service." It was not exactly a non-governmental science and technology enterprise, but an early form of such an enterprise. However, it was not just an ordinary technology consulting service business, as it was aimed at exploring the possibility of developing a high-tech industry like those in Silicon Valley and along Route 128 in Boston. Actually, quite a number of technical people cherished the direction of small business, but they were not sure if it was the right thing to do, as it deviated from the existing system, where all research institutions and enterprises were under government administration. Of course, there were a group of people who opposed it as something like heresy. In 1982, insofar as "economic construction must rely on science and technology, and science and technology must be oriented toward economic construction" was promoted as the general orientation for China's scientific and technological development and the newly appeared non-governmental science and technology business was supported by some leaders, a consensus was reached regarding the development of non-governmental science and technology enterprises. In 1983, the small "Advanced Technology Service" transformed into a non-governmental science and technology economic entity—Huaxia New Technology Development Institute. It was followed by a number of similar companies. In May, 1983, the New Technology Joint Development Center was established. This was a collective-owned company jointly supported by the Chinese Academy of Sciences and the Haidian District Agricultural, Industrial, Commercial Combine Co. In July, the Computer Research Institute of the Chinese Academy of Sciences combined with a financial unit in Haidian to establish the "Kehai Computer Room Technology Development Company," which was the first specialized computer machine room company in China. By 1983, the non-governmental science and technology enterprises on the Electronic Street increased to 11. The period from 1980 to 1983 could be seen as the nurturing stage of the Electronic Street.

The growth stage started in 1984. In April, 1984, a small group of middle-aged college professors, supported by their own college and the Haidian district, set up Haihua New Technology Development Center. In May, an engineer came from the Computer Center of the Chinese Academy of Sciences to start up the now famous Stone Company jointly with the Haidian Sijiqin Commune. In October, a scientist from the Institute of Geophysics of Chinese Academy of Sciences came along to set up the Northern Computer Company jointly with the Zhonguancun Street Collectives Combine. In November, the Xingtong Company was established as a shareholding limited liability company jointly invested in by the Computer Technology Research Institute of the Chinese Academy of Sciences, the Scientific Instruments Factory of the Chinese Academy of Sciences and United Newly Rising Industry General Company with a million yuan investment from each of the three. All of these were backbone enterprises of the Electronic Street. By the end of 1984, the enterprises on the Electronic Street had increased to more than 440; and in 1985, they increased to more than 490. In early 1988, the proposal to establish the Zhongguancun new technology industrial development experimental zone was presented, and in May, 1988, the "Provisional Regulations on Beijing New Technology Industrial Development Experimental Zone" was approved by the State Council. The development of the Electronic Street entered a new stage. By the end of June, 1988, the science and technology enterprises in the area around the Electronic Street had increased to 503, and more than 300 new enterprises had applied to set up on the Electronic Street. After 1992, under the influence of the development of a market economy, the Electronic Street saw more rapid growth; now, there are more than 10,000 science and technology enterprises on the Street. The whole area has been expanded further.

Why can non-governmental science and technology enterprises be developed so quickly and so successfully on Zhongguancun Electronic Street? Zhongguancun has the highest concentration of knowledge in the country, or maybe even in the world. In this small area, there are 50 higher education institutions, including world-famous universities like Beijing University and Qinghua University. There are 41 research institutes under the Chinese Academy of Sciences, 72 research institutes under various central ministries, 27 research institutes under various municipal departments and 267 research institutes affiliated with various universities and colleges. On average, there is one independent research institute per 0.55 kilometer. In the area, there are 80,000 scientific and technical personnel, more than 100,000 university undergraduate and graduate students, 20,000 university graduates and 4,000 graduate students receiving their degree every year. With 23,000,000 books in the libraries and a concentration of research facilities, this area has the best environment for the development of science and technology enterprises.

The Zhongguancun Electronic Street, the first technology intensive area in the country, made quite a few breakthroughs in China's modernized industrial development. First of all, as most of the enterprises in this area were non-governmental enterprises, they did not develop on government support or investment. Most of the enterprises raised their own capital, started on a very small scale and developed at a very high speed with very high rate of return. In 1984, the total business volume was more than 18 million yuan. In 1985, it increased to 140 million yuan. In 1986, it further increased to 350 million yuan. In 1987, it exceeded 900 million yuan. According to a survey of 43 companies, in 1987, the per employee profit was 42,000 yuan. Among them, the Xingtong Company had a per employee business volume of 800,000 yuan and per employee profit of 52,000 yuan; while the Stone Company had a per employee business volume of 760,000 yuan and a per employee profit of 60,000 yuan. In the three years from 1985 to 1987, all the enterprises in this area made a total after-tax profit of more than 78 million yuan, while total fixed assets of all these enterprises were 47 million yuan.

Second, aside from the economic effect, the enterprises on Zhongguancun Electronic Street also achieved a number of technological development results, developed quite a number of new products and advanced the new products to either the domestic or the international level. By the end of 1987, 740 technology results had been achieved, and 556 of them had been developed into products. Among them, 40 products had entered the international market. The artificial intelligence breath monitor and other sensors, as a series of sensor products developed by Xingtong Company, won the only gold medal at the International Invention and Technology Trade Fair held in Germany in 1997.

Third, the characteristics of the R&D of the science and technology enterprises on Zhongguancun Electronic Street were short cycle, with a high application rate and strictly followed market-oriented direction (Table 24).

The experiences of the science and technology enterprises on the Zhongguancun Electronic Street proved that the success of these enterprises is mainly due to the emergence of a new operational mechanism. The new operational mechanism is embodied in the following aspects:

To make decisions on one's own accord. All science and technology enterprises on the Electronic Street, notwithstanding their ownership, were

Table 24
Transformation Cycle of R&D Results: Statistics of 54 Companies

Project Background	Within 6 Months	Between 6 Months and a Year	1 to 2 Years	Above 2 Years
Projects Brought from Parent Institute	13	14	21	23
Independently Developed	105	64	44	10
Developed by Entrustment	58	61	16	3
Cooperative Development	32	14	10	28

Source: "Guide to China's Science and Technology Policy."

able to make decisions on their own accord. Part of the State-owned enterprises were under the Chinese Academy of Sciences. However, the Academy didn't interfere with the business operations of the companies. The companies carried out the general manager responsibility system under the leadership of a board of directors. The collectively owned science and technology enterprises, right from the start, had the full power of decision making on issues with respect to personnel, finance, materials and operations. All the development projects were selected and decided on by the enterprises themselves, and all materials were sourced by the enterprises. It was exactly due to making decisions on their own accord that the enterprises could make quick decisions in competition, react very quickly and win the time they needed. Time is the most important factor in competition. For instance, the Stone Company, established in 1984, in the same year developed the M-2034 printer. The Company solved technology problems in only eight days and spent three days in negotiating and signing a contract with the foreign business. The Company took a one month loan of a million yuan from the bank. After the products were turned out, all were sold out in three days. The Company paid off the bank loan thirteen days ahead of schedule and made a profit. Another State-owned enterprise entered into negotiations with the foreign business on the same project earlier than Stone Company did, but the State-owned enterprise had to prepare a lot of reports and reference materials and submit them to the senior authority. The manuscripts of these reports and reference materials weighed about 50 kilograms. The negotiations had taken half a year without making a deal. The failure was due exactly to the lack of independent decision making.

The market-oriented operation. There are two basic points in the operation of the science and technology enterprises on the Electronic Street—to take science and technology as the backbone and to take the market as the orientation. In the competitive market environment, only by viewing the market, studying the market and conforming to the market can the enterprise penetrate the market and develop market share. By practicing the market-oriented operation, science and technology enterprises on the Electronic Street put their whole operation under the direction of market demand. Right from the selection of projects to performance and pricing to packaging, transportation, marketing and after sale service, all links centered around market demand. By so doing, each of the science and technology enterprises has developed its main product, such as the electrostatic resistant floor of the Jinghai Company and the MS series printer of the Stone Company.

The two-way personnel flow mechanism. With respect to personnel management, the science and technology enterprises on Electronic Street carried out a contract system and practiced two-way free selection. The employees had the freedom to select enterprises and jobs. It meant the free flow of personnel, including free flow inside the enterprise. On the other hand, enterprises, including different departments within the enterprises, had the freedom to select their working staff. With the combination of the two aspects, enterprises are able to form an optimal composition of their working staff. Insofar as the rigidity of the original personnel management system remained an influence in personnel management in State-owned enterprises, the brand new personnel management system carried out in non-governmental enterprises had a great attraction for talented people. The new personnel management system of non-governmental science and technology enterprises helped to promote the efficiency of the business operation.

The rational distribution of benefits. Science and technology enterprises obtained good economic benefits. According to a general estimate, the average economic benefits of the companies on the Electronic Street in the few years after their establishment were 10 times that of the average State-owned enterprises. During the initial stage of development, the proportion of the distribution of profits was 5:3:2, that is, 50% development funds, 30% welfare funds and 20% bonus funds. With further development and the growth in the strength of the enterprises, the proportion of development funds increased and welfare and bonus funds decreased. The personal income of the staff members of the enterprises is comprised of three parts: basic salary, duty subsidy (to be decided by the specific duty and professional title) and efficiency salary (bonus). The difference between the highest duty subsidy and the lowest may be a factor of seven, and the difference between the highest bonus and the lowest may be a factor of nine.

The emergence and development of the Zhongguancun Electronic Street had been an important step in both economic reform and scientific and technological reform. As an example for the development of non-governmental enterprises, the vigor of the new management system has been proved. As an example, the small group of science and technology enterprises have broken new paths in making science and technology grow into the economy.

The virtual development of Zhongguancun Electronic Street is far beyond what people had expected. After 1992, the companies on the Electronic Street had more rapid development not only in scale but also in depth. Some of the most successful companies developed their manufacturing bases in other high and new technology development zones in Beijing and have even reached out to coastline areas, and sales of their products have grown to billions upon billions of yuan.

NOTE

1. Zhongguancun is the near suburb of Beijing, located at the western part of the city.

19

Policy on Sustainable Development

The construction of a modernized China is being carried out under the conditions of large population, limited resources per capita and a comparatively low level of economic and scientific and technological development. Since the 1970s, with growth in population, economic development and the enhancement of people's consumption level, heavier pressure has been imposed on a country that has suffered from a comparative scarcity of per capita resources and fragile environment. Therefore, it is a major issue to select the path of development closely related to the survival of both the contemporary Chinese people and their descendants.

The only choice is to take the path of sustainable development. This means that economic development must be kept in balance with the reasonable exploitation of resources and the protection of the environment, or in other words, economic development should not be realized at the expense of resources and the sacrifice of the environment. There is a saying in China: "Don't eat everything up including those which belong to your descendants."

However, the path taken by history has not always been a straightforward one. People must learn from their own experiences, including the bad experiences. High tuition has to be paid for people to reach wise judgment regarding their own survival. Chinese people have indeed paid a high tuition in order to gain the experiences and reach the current policy for sustainable development.

OVEREXPLOITATION OF WATER RESOURCES

In the past, people took water resources for granted and used them as if there would be a limitless endowment by Nature. Only in the past couple of decades have people found that Nature does not endow them with resources as generously as they expected.

According to scientists, the water resources utilized by mankind can be only 12,000 cubic kilometers per year. According to a World Bank report published in 1995, 40% of the world population has been confronted with crises derived from the shortage of water resources. The vice president of the World Bank pointed out that in 1996 there were 90 million newborn babies, and according to the population growth rate, after another century the drinking water in the world will be extremely limited.

The rapid growth of world population makes the per capita water resources rate decrease drastically. In the beginning of twentieth century, the world population was 1.6 billion, and now it is 6 billion. This means that the per capita fresh water resources rate is lowered to one fourth that at the beginning of the century. In addition, industrial and agricultural development has changed the consumption structure of water resources. According to statistics provided by experts, since the 1970s, the expansion of agricultural irrigation had caused water used in agriculture to increase by 70%, while industrial development has made that used by industry rise by 20%. The rest of the water was consumed by the rural and urban dwellers. This means that the proportion of total consumption of fresh water resources used for people's daily life has consistently dropped with ongoing industrial and agricultural development, although the urban and rural population has consistently increased. The expanding gap between the demand and supply of water resources certainly brings about serious shortages of water in certain areas, and the situation is aggravated each day.

The situation of water resources in China is not promising. China's per capita water resources rate is less than one fourth that of the world. This has something to do with the historical background. Building dikes in lakes to claim farmland was a traditional way of expanding farmland in China. Before the Tang and Song dynasties, there were a number of large lakes in North China. The famous lake beside the Liang Shan Mountain was a large one in Shandong Province. In the late Ming Dynasty and early Qing Dynasty, large scale projects to build dams to claim farmland had been carried out on the lake. By the late nineteenth century, the total area of this lake had shrunk to one tenth of that in the early Qing Dynasty. The Ji Zhong Plain (located in central Hebei Province) was originally a large lake called "Wen An Wa," with a total area of 1,250 square kilometers. It was entirely due to land claiming efforts that the lake entirely disappeared and the whole area became a vast plain.

The land claiming efforts continue today. Consequently, by 1972, the Dong Ting Lake, the largest freshwater lake in China, shrank to 75% of its size in 1949. In 1980, Hong Hu Lake shrank to 60% of its size in 1957. There were once a large number of lakes scattered over the vast plain along the middle reach of Yangtze River, which made the area known as "a place of a thousand lakes." In recent years, more than 6,000 square kilometers of lake area have been dammed and claimed as farmland, which leaves a total lake area of merely 2,300 square kilometers.

In China, more than 300 cities are short of water, and one fourth of these suffer serious shortages. Extracting underground water has been a way to make up the shortage. However, the natural compensation of underground water is limited, and over-intensified extraction will cause the groundwater level to draw down by a big margin and lead to earth subsidence. For instance, in Shanghai, due to over-intensified extraction, the earth subsidence reached five meters by 1965. In the 1970s, by the adoption of a series of measures including limiting water consumption and artificial recharge, the earth subsidence was effectively brought under control. In Jinan, the provincial capital of Shandong Province, more than 100 springs and fountains are located in an area of four square kilometers inside and outside the old town. In recent years, big changes have taken place in most of these springs and fountains. During the dry season of spring every year, many of these springs and fountains are dry, mainly due to the over-intensified extraction of underground water. As extraction is 1.5 million cubic meters per day, while natural compensation is only 1.32 million cubic meters per day, from 1958 to 1980 the ground water level has gone down about 5 meters.

The distribution of water resources in China is uneven. North China is short of water. The cities of Beijing and Tianjin, as well as Shanxi and Hebei provinces, suffer serious water shortages. In 1984, the actual water supply of the whole region was 31.8 billion cubic meters, including 11.3 billion cubic meters of surface water and 20.5 billion cubic meters of groundwater, which accounted for 36% and 64%, respectively. To approach this from the composition of water consumption, the water used for industry accounted for 26%; that used for urban areas, 5%; that used for agriculture, 79%. The shortage of water during the average drought year was 7.8 billion cubic meters; Beijing had a shortage of 700 million cubic meters; Tianjin had a shortage of 1.19 billion cubic meters; Shanxi had a shortage of 1.2 billion cubic meters and Hebei had a shortage of 4.7 billion cubic meters.

Other problems have to do with malfunctions in water management. The majority of consumed water is for agricultural irrigation. In North China, irrigation by extracting groundwater is very popular. This has caused serious over-intensified extraction of groundwater. In the plains and mountain valleys with dense population and developed industry and agriculture, accumulated over-extraction reached 24.6 billion cubic meters. The groundwater buried beneath the plain in front of the mountains has dropped to below 10 meters. The earth subsidence thus derived is very popular. The ground surface of some places in Tianjin, Tanggu and Hangu has been drawn down to below sea level.

On the other hand, many industries did not make serious efforts in technological renovation for saving water. The multiple functions of water use and water recycling have stayed at a low rate. There is a serious waste in water consumption. In industry, the water consumption per 10,000 yuan industrial output value was 446 cubic meters. In agriculture, the utilization index of the irrigation channel systems was about 0.5.

THE TENSION OF LAND RESOURCES

China has relatively limited land resources in connection with her large population. The shortage of land resources is aggravated by the deterioration of the ecological environment caused by various manmade reasons. The severe destruction of forests and grasslands and the desertification process have caused serious loss of arable lands.

China has the lowest vegetation rate in the world. According to statistics in the late 1970s, it ranked 120th in the world. In recent years, even with efforts in reforestation, the vegetation rate has risen only to 14%. The low vegetation rate is one of the major reasons for abnormal climax conditions. The loss of water and erosion of soil as well as the increasing natural disasters of wind and sand storm, flood and drought, have resulted in a lower level of output in agriculture and animal husbandry.

The per capita national income of Guizhou ranks last in the country. Damage to the ecological environment is one of the major reasons for poverty in this province. Even in the 1940s, the vegetation rate in this province was around 30%. However, it dropped below 10% in the 1950s. With efforts spent on reforestation, it has come up to 14.8% in the 1990s, but the vegetation rate is still low. The decrease in the forest area makes rain water evaporate faster, and as a result the number of cloudy and rainy days has decreased. In the meantime, due to the rapid increase in population in this province and the relative decrease of per capita land area, people of this province have spared no efforts in exploiting the originally vegetated lands, which has led to a serious lack of water and the erosion of soil. Up to 1987, the total area of water and soil loss reached 76,700 square kilometers, which accounted for 43.5% of the total territory of Guizhou Province. In 1993, clay flow, land slides, earth cracks and ground subsidence caused by the lack of water and soil accounted for 1,245 cases in 45 counties, which destroyed more than 40,000 ha of farmland.

Originally, there were about 3.3 billion mu of grasslands in China. It is estimated that altogether more than 700 million mu of grassland has been lost to sand, alkalization and degradation.

In the past decade, the area of desertification have expanded to 2,100 square kilometers per year on average. It is estimated that the total area of desert has increased by 300 million mu since the founding of the People's Republic. To a great extent, the destruction of grasslands for the reclamation of farmland has accelerated the process of desertification. In certain parts of Inner Mongolia, there were repeated efforts at opening farmland on grasslands, which made the total area of 18 million mu of grasslands become vast expanses of desert. In certain parts of Hebei Province, an excellent pastoral area with a total grassland area as large as 12 million mu has been vastly exploited, while the remaining odd pieces of grasslands have been seriously degraded by over-rearing and the number of days of windstorms and sandstorms has strikingly increased.

Currently, the deserts, Gobi, desertification lands and lands suffering from sand wind altogether come up to 1.53 million square kilometers, accounting for 15.9% of the total territory of the country.

ENVIRONMENTAL POLLUTION

The loss of resources is fatal for the survival of a country with a large population and a relatively low level of resources per capita. The situation is aggravated by environmental pollution. With the rapid economic development in China, the environment has progressively deteriorated.

A general evaluation of China's environmental situation was made in 1995 by China's State Environment Administration. As a general approach to the environmental situation in China, the Administration pointed out that, as a result of the country's large population base, increased economic growth and lower technological and management levels, as well as a serious waste of resources, the environmental pollution and damage to the ecology which previously centered around urban areas was expanding and spreading to rural areas. The scope of the ecological damage was expanding, too. The issue of the environment has

Table 25
Sewage Displacement Throughout China, 1990 and 1993 (Unit: Thousand Tons)

	1990	1993
Total Volume of Sewage Displacement	35,400,000	35,560,000
Industrial Sewage Displacement	24,900,000	21,950,000
Chemical Oxygen Consumption	7,080	6,220
Heavy Metal Displacement	2.189	1.621
Arsenic Displacement	1.226	907
Cyanogen Compound Displacement	3,891	2,480
Volatile Phenol Displacement	9,325	4,996

Source: China State Environmental Bulletin, announced by the State Environment Administration.

grown into an important factor that will restrain economic development and impact the health of human beings.

The statistics show that the displacement of water sewage, waste gas and solid sewage has increased and that environmental pollution has been aggravated in the past two decades.

WATER POLLUTION

According to observations as early as 1980, 27 main rivers in China, including the Yangtze River, Yellow River, Huai River, Pearl River, Xiangjiang River, and Songhuajiang River, were polluted to various degrees. Each year, about 10 billion tons of sewage were discharged into the basin of the Yangtze River. The pollution was especially serious in sections of the river that flow through Chongqing, Wuhan, Nanjing and Shanghai. Some 2.3 million tons of industrial and domestic sewage were discharged into the Huangpu River, a branch of the Yangtze River, each day.

According to the Bulletin issued by China's State Environment Administration on China's environment situation, a comparison can be made between the sewage displacement in 1990 and that in 1993 (Table 25).

According to the statement of the State Environment Administration in 1996, in the previous year, except for part of inland rivers and large scale reservoirs, the pollution of rivers, lakes and reservoirs throughout the country tended to be more serious than before, and the pollution of water adjacent to industrially developed cities and towns was especially serious. Organic pollution was especially serious in the basins of the Yellow River, Pearl River, Huai River, Songhuajiang River, Liao River and Haihe River. In the four sea areas in China, the pollution of the Bohai Sea and the East China Sea was the heaviest. The deterioration in the water quality of offshore water was even more evident. Red tides took place quite often. No fundamental improvement has taken place in the deterioration of the ecological environment in fishing areas. Displacement of industrial and domestic sewage is the source of water pollution. According to the survey in 1994, the pollution of 110 key rivers in 100 major industrial cities by hydrogen nitride, urea, inactive phenol, colibacillus, and the like was aggravated. In addition, a number of localities had difficulties in water consumption due to insufficient sewage treatment equipment, the inability of sewage treatment and the lower recycling rate of industrial utilization of water, as well as the high water consumption in the paper making, fiber, iron and steel industries. The water of quite a few branches of the Yellow River could not be used for irrigation because of serious pollution. The expansion of the use of agricultural chemicals, though it helped to improve the output of crops, had brought about serious water pollution. In the past two decades, while China saw a constant increase in the use of chemical fertilizers, agricultural plastic films and pesticides, two thirds of the rivers in the country and more than 10 million ha of farmland were polluted. Some places tried to irrigate farmland with industrial and domestic sewage. Insofar as the sewage contained nutrition required by crops, especially nitrides, phosphorus and potassium, it did increase the output of crops. However, this sewage, especially the industrial sewage, also contained various kinds of poisonous and hazardous substances, which polluted the soil. In the sewage irrigated areas in both Shenyang and Shanghai, large areas of paddy fields suffered cadmium pollution. People were reminded that irrigating with sewage was not a good way to raise the output of crops; if it was necessary to use sewage to irrigate, certain steps should be taken to analyze the chemical content of the sewage and to get rid of hazardous elements through treatment.

There has been serious pollution of lakes. The Chaohu Lake in Anhui, Dian Chi Lake in Yunnan, Taihu Lake and Yangchen Lake in Jiangsu and Jingpohu Lake in Helongjiang are all large lakes in China that have been polluted. Various poisonous and hazardous substances have been discovered through analysis.

In the mid-1990s, several hundred million cubic meters of poisonous and hazardous sewage were discharged into the upper reaches of the Huai River, which deprived local people of water sources. Large numbers of industries were forced to stop production, tap water was not suitable for drinking, and the farmlands lacked water for irrigation. In one county, a couple of hundred thousand people lived on water supplied by the PLA's water delivery wagon and the emergency water pipeline. Later, in the middle and lower reaches of Huai River, large areas of water pollution made it difficult for local people to get drinking water. This directly impacted the living and production level of people living around the lower reaches of the Huai River. The situation had drawn the attention of the government. The central government and the local governments at various levels of Jiangsu, Anhui, Shandong and Henan provinces took a number of measures to tackle and prevent the pollution of the Huai River. This included prohibiting further displacement of industrial sewage by industries and closing down a number of small township industries that were sources of heavy pollution.

WASTE GAS

China is a country that uses coal as main energy resource. Some 70% of smoke dust and 90% of carbon dioxide come from coal combustion, which causes serious air pollution in industrial areas and densely populated cities throughout the country. According to the experts, China's air pollution is a result of coal smoke, and the heaviest hazards are dust and acid rain. Acid rain is mainly distributed south of the Yangtze River and east of the Qinghai-Tibet plateau and Sichuan Basin. Central China suffers the heaviest acid rain pollution. In its central part, the average pH value of acid rain is below 4.0, and the frequency of acid rain is above 80%.

Because of the development of urban transportation, pollution from automobile emissions is aggravated, and nitride oxide has grown into the first wintertime pollutant in Guangzhou and Beijing. This shows that the type of air pollution in the very large cities is in the process of transformation (Table 26).

The industrial waste gas discharged into the air contains many poisonous substances. With the effect of gravity or along with rainfall, these substances fall to the surface and thus pollute the soil, causing serious soil pollution in the neighborhoods of large scale industries or mines. For instance, in Baotou, an area of more than 10,000 kilometers was polluted by fluorine in the late 1970s. High fluorine contents were found in air, water, soil, grasslands and crops. More than a million people and a million cattle suffered from the hazards. According to monitoring, waste gas with fluorine hydrogen content displaced by a larger aluminum factory could amount to more than 700 million cubic meters, and within one kilometer of the factory, the fluorine content in air per kilometer could be as high as 180 kilograms. Even within 4 kilometers, the fluorine content in the air still exceeded the standard by about a factor

Table 26
Waste Gas Displacement in China

	1990	1993
Waste Gas Displacement (Excluding Township Industry)	8.5 trillion standard cm.	11 trillion standard cm.
Smoke Dust	13.24 million tons	14.16 million tons
Sulfur Dioxide	14.95 million tons	17.9 million tons
Total Suspended Particles	367 microns/cm. (average value/day)	108/815 microns/cm.
Nitride Oxide	42 microns/cm. (average value/day)	10/815 microns/cm.

Source: China State Environment Bulletin, announced by the State Environment Administration.

of ten. People found that polluted air was another main source of soil pollution, in addition to polluted water.

Pollution from the displacement of industrial sewage, waste gas and solid waste by township industries grew rapidly with the development of township industry. Though the proportion of displacement of industrial waste by township industry was lower, the average concentration of the displacement of main pollutants was quite high. For instance, the concentration of volatile phenol discharged by township industry was 9.9 times that of urban industry. Some township industries, such as paper making, sulfur removal, coking, tannery and dye processing, caused the heaviest pollution. The small township paper making industries using wheat and rice stalks as raw materials adopted the traditional limestone technique. The technical equipment remained at the level of the 1940s and 1950s. Pollution treatment facilities were not available in 70% of the 7,000 small township paper making industries. The polluted sewage directly discharged by these paper making industries exceeded the standard by three or four times. A small paper making factory at an annual capacity of 700 tons using the limestone technique usually caused water pollution of several kilometers. In 1990, the industrial sewage discharged by township paper making industries accounted for 43.8% of that of the entire township industry. The serious pollution of the Huai River was, to a great extent, derived from pollution from small paper making industries. There were about 7,000 small township paper making industries throughout the country. After the pollution of Huai River, the government issued an order to close down all small township paper industries with an annual capacity below 5,000 tons.

The township sulfur removal industry also produced serious pollution.

In China, 70% of sulfur demands were produced by the local traditional technique. The technique, which had been adopted by most of the township sulfur removal industries, caused a serious waste of resources, sacrificed the environment and was hazardous to people's health. In 1984, in Zhengxiong County of Yunnan Province, where scorched earth covered an area as large as 22,500 mu, within a perimeter of 20 square kilometers, none of the trees survived. In order to overcome the serious hazard caused by the traditional technique, a new technique, "Yunhuan 09—closed loop sulfur removal" was developed. The new technique, appraised by the experts as "appropriate renewing technology," brought concentrations of sulfur dioxide and hydrogen sulphide around the sulfur removal area down to the air quality standard limit. Since then, the lotus has blossomed and the birds have come back, and in the scorched earth zone vegetation has recovered on 5,700 mu of land.

In order to control pollution made by township industries, government orders were issued stipulating that before the end of September, 1996, all paper making industries at an annual capacity below 5,000 tons, tannery industries at an annual capacity below 30,000 pieces of hides, dye processing factories at an annual capacity below 500 tons and the coking and sulfur removal industries that use backward techniques should be closed down. Also, township industries using local traditional techniques to produce arsenic, mercury, lead or zinc or for oil refinery or gold dressing or to make pesticides, engage in bleaching and dyeing or electroplating, or to manufacture asbestos products or radioactive products should be closed down.

GOVERNMENT POLICY ON SUSTAINABLE DEVELOPMENT AND COUNTERMEASURES

The deteriorating situation in resources and environment has drawn serious attention from the Chinese government. A whole set of strategies, policies and countermeasures have been worked out by the government. The scientific community has also done earnest study on the issue.

It is acknowledged by scientists that Asian countries, especially the East Asian and Southeast Asian countries, have a rapid growth rate, high population density and a fast increase in energy and resources consumption and that therefore the problems of resources and environment are massive. The rapid economic development aggravates damage to resources and the environment, such as the shrinking of tropical forests, the loss of water, the erosion of soil, the desertification of lands and the pollution of air, rivers and oceans. The Asian acid rain area, one of the three largest acid rain areas in the world, is continuously expanding. With the deterioration of the environment, there is serious damage to land, water, mineral resources, forests and biological resources. All these factors have threatened and hindered sustainable development in Asia.

On the other hand, the United States and Europe had earlier economic development. As during the early stages of their economic development, there were plenty of resources and a smaller global population, the problem of resources and environment was not as outstanding as it is today. The problem is that the developed countries, with a total population of a billion during their past economic development, consumed 80% of the energy and resources and thus caused unprecedented damage to the global environment. Should Asian countries with a total population of 3.2 billion step up and take the traditional development model and pursue high speed economic growth at the expense of a high consumption of energy and resources, it would be impossible to reach the goal of economic development with the exhausted resources of today. A new path of economic development must be taken by Asian countries, including China.

The options for China are either to gain fast economic growth at the expense of high consumption and even a serious waste of resources and the sacrifice of the environment or to closely integrate economic development with the conservation of resources and the protection of the environment and strive for the coordinated development of economy, environment and society. It requires a well-settled relationship between near-term benefits and the long range interest of the people of the country.

At the Fourth National Environment Protection Conference, Jiang Zeming made a statement that gave a clear-cut answer to the question. He indicated that implementing the strategy of sustainable development should be grasped persistently as a major event in the construction of socialist modernization. He pointed out:

> Economic development should be considered in an overall way together with issues of population, environment and resources. We should not only get the current development well-arranged, but also consider our descendants, to create a better condition for the future development. What we should not do is to take the path of pollution first and correcting later, or even to eat up what's left by our ancestors and break off the survival of our descendants.

According to Jiang, a few major aspects merit our attention:

> At first, insist on saving water, land, energy, materials, grain and other kinds of resources. High yields, fine quality, high efficiency and low consumption should be realized in agriculture. In industry, attention should be paid to quality, low consumption and benefits. Tertiary industry should be developed in coordination with first and second industries. Secondly, continue to bring population growth under control and to improve the overall quality of population. Thirdly,

the consumption structure should be reasonable and the consumption pattern should favor of protecting the environment and resources. High consumption divorced from the level of production which causes the waste of resources, should not be developed. Fourthly, further the publicity and education of environment protection and to strengthen the sense of consciously protecting the ecological environment. Fifthly, resolutely keep down and turn around the tendency of the destruction of resources and deterioration of the ecological environment in some areas.

He also stressed:

Our economic and social development should be based on optimizing industrial composition and coordinating between economy, society and environment. As a matter of fact, should the economic growth be based on blindly expanding the investment and arbitrarily developing new setups, the faster the speed of growth, the heavier the waste of resources, the more serious the pollution of the environment and damage to ecology, and the lower the capability for sustainable development.

Jiang's statement expressed the government's determination to coordinate the development of economy, society and environment and to adopt a policy of sustainable development, which, as repeatedly announced by the government, has been selected as a basic development strategy and basic State policy. It was in this regard that China attended the U.N. conference on environment and development in 1992. China paid special attention to the conference. In accordance with the demands raised by the Twenty-First Century Agenda adopted by the conference, China's Twenty-First Century Agenda—White Paper Book on Population, Environment and Development in the Twenty-First Century was formulated and approved by the State Council. Aside from the Agenda, the Chinese government has taken an important step to cover the sustainable development strategy and the measures for its implementation into the Ninth Five Year Plan for the National Economic and Social Development and the Outline of the Long Range Goal up to 2010 of the PRC. The Outline has independent sections devoted to goals, tasks and targets for conserving and developing resources within the national territory and environmental and ecological protection as well.

The Agenda covers both the strategy and specific countermeasures for sustained development in China. Following are highlights of a few major approaches made by the Agenda in this regard.

China's Strategy and Countermeasures of Sustainable Development

The Agenda provides the framework of a comprehensive, long standing and incremental strategy for sustainable development and corresponding countermeasures. This is a new starting point for China to step into the twenty-first century and to strive for a bright future.

The Agenda points out that development is the premise of sustainable development, and it is necessary to satisfy the basic demands of contemporary people, but not to endanger the capability of descendants to satisfy their own demands. China must maintain a faster speed of economic growth and gradually improve the quality of the growth; strive for the sustainable development of society; strengthen environmental protection and adapt economic and social development to the loading capacity of resources and environment and thus gradually realize the coordinated development of population, economy, society, resources and environment in China.

According to the Agenda, the main countermeasures for realizing sustainable development include taking economic construction as the center, deepening reform and opening up and establishing and improving the socialist market economic system; further building up the capability and improving the economic, social and legal systems and the comprehensive decision making mechanism of sustainable development; to carry out family planning, to control the size of population, to improve the quality of the population as well as the composition of population; to promote sustainable agricultural technology by way of adapting measures to specific localities; to readjust industrial composition and layout, to carry out clean production and to extend the comprehensive rational utilization of resources; to develop clean coal technology; to make efforts to develop renewable and clean energy; to accelerate the improvement of urban and rural dwellers' housing environment; to implement major projects on environment pollution control; to earnestly perform various agreements on the global environment in which China has participated and to make unremitting efforts to contribute to the solution of global environment problems.

Regarding the major strategic actions of sustainable development, the Agenda points out: to establish a coordinated management organization for sustainable development; to promote the protection of resources and environment by adopting economic measures; to study how to convert the factors of natural resources and environment into the national economic accounting system; to make the protection of natural resources and environment a basic function of the government at various levels; to gradually transform the prevention and control of pollution from concentration control into total volume control through prevention beforehand rather than tackling it afterward; and to strengthen the publicity and education of sustainable development, so as to enhance the sense of sustainable development of the whole nation.

Regarding international cooperation on sustainable development, the Agenda points out that Chinese government takes the promotion of international cooperation on sustainable development as an important component of China's reform and opening up. The Agenda points out that China recognizes quite well her important responsibility in global sustainable development and environment protection and will take part in various kinds of international efforts to protect the global ecological environment and pursue the sustainable development of mankind with a strong sense of historical responsibility, mainly relying on her own forces, with an active, earnest and responsible attitude. In the meantime, China appeals to the international community to provide China with the necessary support of funds and technology, so as to help China step onto the path of sustainable development as soon as possible.

Looking back at recent history regarding the evolving approach to the relationship between the reproduction of human beings, the exploitation of resources and the deterioration of the environment, people may find that great progress has been made since the 1970s, when a pessimistic view prevailed about the future of population growth and the limit of resources. Now, eventually, the idea of sustainable development provides long range coordinated development between population, resources and environment. Of course, it has a long way to go to attain the ultimate goal.

20

Epilogue

With the development of reform and opening up in the past two decades, China has seen rapid development in science and technology. Science and technology policy does play a role in directing science and technology to develop along the track of serving economic development. In order to let science and technology play a role in promoting economic development, efforts have been made to overcome the separation of scientific research from production. A whole new mechanism—the new science and technology management system—has been developed to integrate science and technology with industrial and agricultural production. The results are noteworthy.

In the light of China's opening up policy, China's scientific and technological development has taken advantage of opening up to the outside world. After 1978, the decisive year for opening up, scientists and engineers took the lead in breaking up the closed loop system and looking around the world's advanced level of scientific and technological development. Since then, international exchange and cooperation have been a priority in China's science and technology policy. Technology acquisition has played an important role in bringing China's technology to new levels.

In the 1990s, with further reform and opening up, the country has both achieved fast economic growth and been confronted with new challenges. The country has entered into a new phase of development, and science and technology must play an even more important role in transforming the economic growth model. More intensified international competition has made it inevitable that China needs to upgrade her industrial composition. High-tech R&D and high-tech industrialization have been given attention in the industrial sector, and innovation strategy has replaced the original follow-up pursuit in high-tech development. In order to prepare the reserve strength of scientific and technological development, a new emphasis has been placed on basic research. Obviously, China's science and technology policy is furthered with the changing situations.

Nevertheless, science and technology policy is unable to play the role all alone. It can be effective only when it is implemented in a suitable socioeconomic context. As long as China is in a transition stage, though the ultimate goal of policies in various sectors are the same and interrelated, the development of practices in different sectors is often uneven. The attempt to push forward the technological progress in large and medium-sized industries has been repeatedly frustrated. The situation necessitates the furtherance of coordinated efforts in both economic and scientific and technological reform.

In addition, sustainable development requires the further development of science and technology policy. New decrees, new norms and new standards should be worked out to gear production to the new concept of sustainability and to match technological development with resource conservation and environmental protection. How to fit scientific and technological development to the perspective of the coordination of three productions, namely, material production, the reproduction of human beings and environmental production, is the task of science and technology policy. Though it is a long range goal for the country, the policy should integrate near-term development with long range goals. In order to bring the country into prosperity, a long and arduous route must be taken, and science and technology should play a role in light of the general development strategy of the country.

References

Bei Yiyan. "Open Policy Boosts Technical Growth." *Beijing Review*, Vol. 30, No. 39, September 28, 1997.

Chen Jinhua. "Grasp the Opportunity and Revitalize the Pillar Industries." *People's Daily*, March 29, 1994.

Chen Jixi. "Universities Cooperate with Enterprises to Promote Scientific Research and Economic Development." *China Science and Technology Forum*, No. 2, 1997.

Chen Siwei. "Prospects of China's Technology Research Policy in the Nineties." *Liao Wang (Survey)*, Overseas Edition, No. 1, 1993.

"China's Twenty First Century Agenda." (White Paper on China's Population, Environment and Development in the 21st Century) (Summary). *People's Daily*, September 19, 1994.

"Decision of the Central Committee of the CPC and the State Council on Accelerating Scientific and Technological Progress." *People's Daily*, May 22, 1995.

"Decision of the Central Committee of the CPC Concerning Reform of the Science and Technology Management System." *People's Daily*, March 20, 1985.

"Decision of the State Council on Deepening the Reform of Scientific and Technological System during the Ninth Five Year Plan." *People's Daily*, October 4, 1995.

Deng Xiaoping. Speech at the Opening Ceremony of the National Science Conference, March 18, 1978.

"Experts Involved with '863' Program Viewing '863' Program." *China Science and Technology Forum*, No. 2, 1996.

Fei Xiaotong. "The Path of the Development of China's Urban and Rural Areas." *China Social Science*, No. 1, 1993.

Fei Xiaotong. "Small Towns, Big Issue." *Liao Wang (Survey)*, Nos. 2, 3, 4, 5, 1984.
Gong Yuzhi. Concerning "Science and Technology Is the First Force of Production." *China's Science Daily*, August 6, 1997.
He Chuanqi. "Initial Consideration on the State Goals of China's Basic Research." *World Science and Technology Research and Development*, No. 1, 1997.
Hu Angang. "Is There an Over-absorption of Foreign Capital in China?" *International Economic Review* Nos. 3–4, 1997.
Hu Qili. "Several Issues on Building China's Information System." *Twenty-First Century*, No. 4, 1997.
"Industrial Catalog Guiding Foreign Investment." Approved by the State Council after Revision in December, 1997. *Beijing Review*, March, 1998.
Jiang Zeming. "Economic and Social Development in Rural Areas." *People's Daily*, July 15, 1996.
Jiang Zeming. Speech at the Fourth National Environmental Protection Conference, July 19, 1996.
Jiang Zeming. Speech at the National Conference on Science and Technology. Xinhua News Agency, May 26, 1995.
Jiang Zeming. "Strengthen Agricultural Foundation, Deepening Reform in Rural Areas, Promote All Around Development." *People's Daily*, July 15, 1996.
Jiang Zeming. "Summing up Experiences, Create a New Prospect in Using Foreign Capital"—Preface of *Basic Knowledge of China's Using Foreign Capital. People's Daily*, May 21, 1995.
"The Law of the Promotion of the Transformation of Scientific and Technological Results of the PRC." Promulgated by the Standing Committee of the People's Congress of PRC. *People's Daily*, May 23, 1996.
Li Peng. "Build a Unified Transport System." *Economic Daily*, August 5, 1997.
Li Peng. "China's Energy Policy." *Qiushi*, No. 11, 1997.
Li Ruihuan. "Several Issues on Agricultural Land." Xinhua News Agency, January 1, 1994.
Liu Jiang. "Agriculture Moves toward Market Economy." *Beijing Review*, April 18–24, 1994.
Luo Guiquan. "Pay Special Attention to the Development and Utilization of Information Resources." *Guangming Daily*, April 7, 1997.
Mao Meng. "From Output Related Responsibility System to Scale Operation." *China's Reform*, No. 2, 1994.
"The Ninth Five Year Plan for National Economic and Social Development and the Outline of the Long Range Goal up to 2010 of the PRC." Approved by the Fourth Session of the Eighth National People's Congress. *People's Daily*, March 20, 1996.
Peng Shutang. "The 'Second Pioneering' of Non-governmental Science and Technology Enterprises and the Issue of Scientific Management." *China Science and Technology Forum*, No. 3, 1996.
Press Office, the State Council of the PRC. "China's Environmental Protection." *People's Daily*, June 5, 1996.
Press Office, the State Council of the PRC. "China's Grain Issue." *People's Daily*, October 25, 1996.
The State Council of the PRC. "Decision on the Several Questions about the

Structural Reform of Science and Technology." Xinhua News Agency, May 19, 1988.

The State Council of the PRC. "Outline of China's Agricultural Development in the Nineties." *People's Daily*, January 19, 1994.

The State Council of the PRC. "Outline of State Industrial Policy in the Nineties." *People's Daily*, June 23, 1994.

"State Outline of Mid-term and Long-term Scientific and Technological Development." *People's Daily*, October 2, 1992.

State Planning Commission. "State Industrial Policy of the Automobile Industry." *People's Daily*, July 4, 1994.

State Science and Technology Commission of PRC. "Guide to China's Science and Technology Policy" White Paper on Science and Technology. Beijing, China: Literature of Science and Technology Press.

State Statistical Bureau. The Statistical Bulletin on the National Economic and Social Development of the PRC in 1992, 1993, 1994, 1995, 1996. *People's Daily*.

Tian Jiyun. "Great Change in China's Rural Areas." *Peasants' Daily*, February 25, 1994.

Wang Yueping and Ge Yuejing. "Consideration on the Strategy for the Readjustment of China's Industrial Formation." *Management World*. No. 2, 1997.

Wang Yougong. "China's Technology Market Is Flourishing." *People's Daily*, February 23, 1988.

Wu Xiang. "New Development and Re-consideration on Wenzhou Model." *Liao Wang (Survey)*, No. 32, August 8, 1994.

Yan Jici. "The Development of China's Scientific Cause in the Past Thirty Years." *Guangming Daily*, November 2, 1979.

Zhang Rong. "Accelerate the Construction of China's Multi-media Communications Network." *Guangming Daily*, August 29, 1997.

Zhao Xiuchen. "Talking about Some Problems on Using Foreign Capital." *Future and Development*, No. 1, 1997.

Zhou Guangzhao. "A Few Points of Strategic Consideration on the Development of China's Science and Technology." *People's Daily*, July 20, 1994.

Zhu Houze. "Twelve Topics on the Development of West China." *Bohai Belt*, May, 1996.

Index

Acid rain, 215, 217
Added value, 56, 59, 80–81, 83
Aeronautic industry, 134
"Agricultural Development Outline of China in the Nineties," 91
Agricultural support system, 96
Agricultural technology extension organization, 72, 99
Agricultural technology extension system, 72, 98
Appropriate technology, 28, 104
Appropriation system, 15, 18–21
Arable land, 85, 87–88
Assimilation and absorption of advanced foreign technology, 13
Automobile industry, 79, 84, 152–53, 155–57, 160

Basic machinery, 140
Basic parts and components, 139–40, 143
Basic technologies, 139–41
BOT (build, operate, transfer), 27, 59, 132
Broadcasting television networks, 149
Building materials industry, 164
Bumper Harvest Program, 95, 99–100
Buyers' market, 25

Capital intensive industry, 59
Capital pool, 107
Chemical fiber industry, 150
Chemical industry, 122, 149–50
China's Twenty-First Century Agenda, 219
Civil aviation system, 129
Closed door policy, 6
Closed loop research system, 20
Coal chemical industry, 122
Combined research and production entity, 12
Commercialization of technology results, 15, 23–24, 111
Commodity production base, 113
Comprehensive national power, 71, 73, 173
Computer industry, 144–46

Computer networks, 149
Conglomeration, 60–61, 106
Construction industry, 139, 162
Consumer electronic products, 46, 109
Contracted responsibility system, 33, 60
Cross linkages, 27
Custom tariff, 58–59, 81
Customized orders, 25

Datong-Qinhuangdao electric railway line, 129
Dayawan Nuclear Power Station, 128
Decision of Industrial Policy, 75
"Decision of the Central Committee of the Communist Party of China on the Reform of Science and Technology Management System," 14, 23
Decision on the Deepening of Scientific and Technological System Reform During the Ninth Five Year Plan, 70
Demand pull, 14, 27, 35
Desertification, 88–90, 93, 212
Development research, 19
Development strategy, 58
Distribution system, 201
Driving force, 69
Durable home appliances, 48

Ecological environment, 88–90, 211, 214, 219, 221
Ecological protection screen, 113
Economies of scale, 81–82, 101
Egalitarian distribution system, 85, 201
Electrical home appliances, 196
Electronic industry, 79, 143–44
Energy composition, 9
Energy industry, 78, 121–22
Enterprise-directed research system, 4
Environmental pollution, 165, 175, 212
Equilibrium of international payments, 80
Export-oriented economy, 56, 83, 105, 116, 186
Export-oriented enterprise, 50
Export-oriented industry, 57, 59, 105, 184
Extension and application rate, 23
External economic environment, 27, 30, 33

Factors of production, 52
"Fengqing Ship Event," 42
First force of production, 8, 68, 182
Foreign Affairs School, 1
Foreign exchange equilibrium, 160
Four modernizations, 8, 15
"Fourteen Points of Opinion on the Current Work of the National Science Research Undertakings," 5

Global corporation, 53, 58
Government-directed research, 4, 70–71
Great Leap Forward, 4–5, 85, 97

Helping the poor, 72, 118
High and new technology industrial development zone, 189
High and new technology industry, 185–86, 188, 196
"High, grouping and outward-oriented," 118
High-tech business incubation center, 187
High-tech industry, 182–83, 185, 187, 195
High Technology Research Program, 173, 175–76, 178–79, 181–84, 186
Highway system, 129, 131
Human pool, 107
Hydraulic power, 127

Immature industrial products, 80
Immature industry, 56, 153
Incubation, 189
Industrial formation, 50, 56–57, 62, 76, 80, 105
Industrial organizational structure, 76, 81–82
Industrial policy, 59, 75–76, 79, 83, 155, 183
Industrial service system, 114

Industry-directed research, 71
Information technology industry, 149
Infrastructure, 76, 78–79, 121, 187
Initial industrial system, 30, 184
Inland river transportation, 133
Innovation strategy, 223
Intellectual property rights, 62, 69, 139, 141, 177, 179
Intensive farming, 101
Intensive operation, 102
Internal demand, 30–31, 35
Investment environment, 44

Jing-Jiu Railway, 130
Joint venture, 43–44, 48, 50, 53–54, 106, 108

Knowledge-based industry, 27

Labor intensive industries, 56, 59, 196
Labor intensive products, 59, 66, 109, 197
Labor intensive projects, 56
Laser High Technology Funds, 179
Laser Youth Funds, 179
"Law of Land Management of the People's Republic of China," 87, 91
Life span, 47, 49
Light industry, 80, 122, 162
Localization, 78, 156, 161

Machinery and electronic industry, 139
Machinery industry, 79, 139–40, 142–43, 153
Macro agriculture, 77, 86
Macro control, 76
Mainstay industries, 76, 139, 79–80, 151, 155, 162
Manufacturing technology, 81–82
Market oriented industry, 184, 206
Materials industry, 162
Micro-electronic technology, 144
Micro-electronics, 79
Microwave industry, 195
Mobility, 21, 37–38, 40
Mobility rate, 39
Modern enterprise system, 67, 157
MPT (Ministry of Post and Telecommunications), 137–38
Multiple crop index, 94
Mutual supplement of superiority, cooperation as, 49

National industrialization, 1
National industry, 1, 55, 58
National information infrastructure, 149
"National Plan on the Prospect of the Development of Science and Technology" (1956–1967), 4
National regional monopoly, 82
National treatment, 59
Nationwide Science, Technology and Personnel Development and Exchange Coordination Network, 23
Ninth Five Year Plan for National Economy and Social Development and the Outline on the Long Range Goal up to 2010 of PRC, 143, 162, 164, 219
Non-governmental science and technology enterprise, 66, 189, 199–203, 206
Non-tariff barriers, 58, 67
Normal cycle, 27
Nuclear power, 127–28, 175

Ocean transportation, 133
Opening policy, 42–43, 58, 67, 186
"Outline of State Industrial Policy During the Nineties," 75–76
"Outline of the Report on the Guidelines for China's Scientific and Technological Development," 10
Output per unit area, 44–45, 101
Output-related contract responsibility system, 85–86, 102

Patent law, 43
Patent right, 160
Pearl River Delta, 105–6, 146
Pearl River Delta township industry, 105
Peer review, 12, 19

People's Commune, 85, 98, 102
Per capita energy consumption, 122
Per capita land area, 87
Per capita land occupancy, 87
Petrochemical industry, 79, 123, 149, 151
Petroleum industry, 124, 126
Planned commodity economy, 42
Policy adjustment, 59
Power industry, 126–27, 175
Preferential treatment, 59
Price mechanism, 79
Price structuring, 79
Primary products, 80
Product economy, 32, 60
Product mix, 9–10, 57, 79, 117
Production factors market, 63
Production Force Promotion Center, 71
Production target, 32
Property right, 60, 117, 163
Property right transfer market, 63
Protectionism, 58, 62
Protectionist policy, 56, 58
"Provisional Regulations on Beijing New Technology Industrial Development Experimental Zone," 203
"Provisional Regulations on Technology Transfer," 22

Qinshan, 128
Quadruplicating of the total agricultural and industrial output, 9
Quantitative economy, 32, 60
Quantitative growth, 65, 68
Quota restriction, 80

Railway system, 129
Red tides, 214
Refinery industry, 149
Regional mainstay industry, 115, 119
Regional protectionism, 140
"Regulations for the Protection of Basic Farmland," 91
Rejuvenation of the nation by science and education, 68
Resource intensive industries, 57
Resource intensive products, 80
Revitalization of national economy, 76

Scale economy, 60, 81, 101, 153, 156, 186
Scale operation, 101–2, 118, 123, 151
Scaling new heights, 166, 171
Scaling New Heights Program, 168
Scaling New Heights projects, 169
Science and technology credit funds, 20
Science and technology development funds, 24
Science-and-technology-oriented businesses, 39
Science-and-technology-oriented enterprise, 189, 199
Science and technology park, 188
Science and technology personnel development tank, 24
"Scientific and Technological Development Plan for the Period from 1963 to 1972," 5
Shareholding cooperative system, 118
Shareholding corporation, 62
Shareholding system, 60–61
Ship-building industry, 132
Shopping basket project, 96
"Short, even and fast," 111, 118
Shortage economy, 32
Socialist market economy, 42
Socioeconomic conditions, 45
Socioeconomic context, 223
Socioeconomic demands, 18
Software industry, 147–48
"Some Stipulations on the Further Promotion of Reform of Scientific and Technological System," 200
Southern Jiangsu, 106
Southern Jiangsu township industry, 105
Spark Program, 18, 95, 104, 110–11, 118–19
Spark Program technology intensive area, 115, 119
Specialized technology society, 99
Spot transactions, 25
State appropriation, 18-19, 21

State engineering centers, 170
State industrial policy, 57, 69, 81, 151
State key laboratories, 169–70
State Mid-term and Long-term Scientific and Technological Development Outline, 167
State-owned enterprise, 60–62, 157, 200–201, 205–6
State-owned industry, 104, 117
State strategy, 58
Sustainable development, 208–9, 217–21
Sustained development, 57, 174
Synthetic fiber industry, 150

Tariff rate, 67
Technology acquisition, 42, 75, 143, 154, 171, 222
Technology acquisition, assimilation and absorption, 140
Technology Extension Center, 71
Technology importation, 44–48
Technology innovation, 12
Technology market, 15, 21–23, 25, 27
Technology policy, 75–76, 129, 135–36, 139, 141
Technology push, 14, 18, 30
Technology renovation, 11, 24, 124, 153, 189
Technology service contract, 25
Technology transfer, 62
Technology transfer contract, 25
Telecommunications industry, 135–36, 138
Telecommunications networks, 149
Television industry, 194
Tertiary industry, 76–77
Textile industry, 80, 150–51
Thermal power, 127
Three Gorges Hydraulic Power Station, 126–27
Torch Program, 28, 184, 186–88, 191
Township industry, 27–28, 104–7, 108–10, 185–87, 194–96, 216–17
Trade barrier, 56
Trademark right, 160
Transfer of technology, 22, 44–45
Transferee, 62
Transferor, 62
Transformation of economic growth model, 67
Transportation system, 128–29
TV program transmission and coverage network, 47

UNICOM (United Communications Corp.), 137
Unified purchasing and selling system, 86
United Technology Market Development Group of China, 24
Urbanization, 83, 87

Vegetation rate, 66, 211
Vertical control, 62
Vertical system, 15, 18
Volume production, 140, 153, 156, 184, 193

Water saving agriculture, 94
Water transportation system, 129
Wenzhou, 106–8
Wenzhou township industry, 105
World Trade Organization, 58, 66

Xiaolangdi Dam, 127

Zhongguancun Electronic Street, 202–4, 207

About the Author

Q. Y. YU was the director of a technology development institution in the People's Republic of China prior to his retirement. His knowledge of the science and technology systems in China and in the United States has been built upon decades of practical experience in technology exchange and technology transfer. While he has published numerous articles in Chinese, this is Mr. Yu's first book in English. He lives in China but has traveled often to the United States, and is former visiting scholar at the Georgia Institute of Technology, associated with the Sam Nunn School of International Affairs and the GT Center for International Business Education and Research.

ISBN 1-56720-332-9

90000>

EAN

9 781567 203325

HARDCOVER BAR CODE